SB 255 .A47 1999
Advances in hemp research

Paolo Ranalli, PhD
Editor

Advances in Hemp Research

Pre-publication
REVIEWS,
COMMENTARIES,
EVALUATIONS . . .

"*Advances in Hemp Research* describes the state of the art of agronomic hemp research worldwide. The book is a fascinating compendium of recent scientific research on cannabinoids, and the botony, breeding, and growing of hemp. It allows us to move beyond the enthusiastic hemp 'fiction' that has prevailed in recent years, so that we can finally discuss the real facts about the plant, with informed debate based (at last) on a top-quality reference work. It is remarkable to find that so many 'aggressors' find *Cannabis* just as attractive as people do, and yet have still not been able to contain the plant."

Christian R. Vogl
*Assistant Professor,
Institute of Organic Farming,
University for Agricultural Sciences,
Vienna, Austria*

More pre-publication
REVIEWS, COMMENTARIES, EVALUATIONS . . .

"Among the large number of English publications on the subject of hemp, this compilation stands out as one of the few written by researchers/practitioners who actually have—sometimes decade-long—firsthand experience with hemp research and implementation. As a result, it provides a well-founded evaluation of several of the aspects relevant to the successful cultivation of *Cannabis,* such as botany and agronomy. Yet, it also touches upon several of the aspects relevant to the real challenge for the crop, i.e., establishing its acceptability to regulatory agencies and, in particular, the suitability of its products for modern markets. While it does not provide exhaustive coverage of the myriad of potential hemp products, it focuses, in an exemplary fashion, on two particularly promising areas, the conversion of hemp fiber into pulp and paper, and the use of its nutritious and tasty seeds and oil for food."

Gero Leson, DEnv
*Leson Environmental Consulting,
Berkeley, CA*

"This book provides an extensive review of recent advances in hemp science and places these advances within the context of existing knowledge and understanding. A broad range of topics is covered including: botany, phytochemistry, methods for measuring THC, agronomy, physiology, diseases and pests, germplasm resources, pulping, and hemp seed as a food source.

Having recently completed a three-year research project on hemp, I can greatly appreciate the value of such text as a source of detailed and credible scientific information and comprehensive reference lists. The rigorous scientific approach adopted in the work reported in this reference will assist to dispel or qualify many hemp 'myths' and contribute to improving the credibility of hemp as an alternative crop option, particularly in nontraditonal hemp-growing regions where lack of quality, accesible literature is often an obstacle to policy change and industry development."

Shawn Lisson, PhD
*Crop and Soil Modeller
in Sugarcane Systems,
CSIRO Tropical Agriculture,
Queensland, Australia*

More pre-publication
REVIEWS, COMMENTARIES, EVALUATIONS . . .

"**A***dvances in Hemp Research* combines all the latest achievements in *Cannabis* study. A must for biologists, physiologists, and every hemp grower, this book is full of scientific information on hemp varieties, the technology of hemp growing, and the use of hemp products."

Sitnik Vasilis
*Institute of Fiber Crops,
Ukraine*

Food Products Press
An Imprint of The Haworth Press, Inc.

NOTES FOR PROFESSIONAL LIBRARIANS AND LIBRARY USERS

This is an original book title published by Food Products Press, an imprint of The Haworth Press, Inc. Unless otherwise noted in specific chapters with attribution, materials in this book have not been previously published elsewhere in any format or language.

CONSERVATION AND PRESERVATION NOTES

All books published by The Haworth Press, Inc. and its imprints are printed on certified pH neutral, acid free book grade paper. This paper meets the minimum requirements of American National Standard for Information Sciences–Permanence of Paper for Printed Material, ANSI Z39.48-1984.

Advances in Hemp Research

FOOD PRODUCTS PRESS
Crop Science
Amarjit S. Basra, PhD
Senior Editor

New, Recent, and Forthcoming Titles of Related Interest:

Dictionary of Plant Genetics and Molecular Biology by Gurbachan S. Miglani

Advances in Hemp Research by Paolo Ranalli

Wheat: Ecology and Physiology of Yield Determination by Emilio H. Satorre and Gustavo A. Slafer

Mineral Nutrition of Crops: Fundamental Mechanisms and Implications by Zdenko Rengel

Conservation Tillage in U.S. Agriculture: Environmental, Economic, and Policy Issues by Noel D. Uri

Advances in Hemp Research

Paolo Ranalli, PhD
Editor

Food Products Press
An Imprint of The Haworth Press, Inc.
New York • London

Published by

Food Products Press, an imprint of The Haworth Press, Inc., 10 Alice Street, Binghamton, NY 13904-1580

© 1999 by The Haworth Press, Inc. All rights reserved. No part of this work may be reproduced or utilized in any form or by any means, electronic or mechanical, including photocopying, microfilm, and recording, or by any information storage and retrieval system, without permission in writing from the publisher. Printed in the United States of America.

Cover design by Jennifer M. Gaska.

Library of Congress Cataloging-in-Publication Data

Advances in hemp research / Paolo Ranalli, editor.
    p. cm.
Includes bibliographical references and index.
ISBN 1-56022-872-5 (alk. paper)
    1. Hemp. 2. Cannabis. I. Ranalli, Paolo.
SB255.A47 1998
677'.12—dc21
    98-12986
    CIP

CONTENTS

About the Editor	ix
Contributors	xi
Preface	xiii

Chapter 1. Botany of the Genus *Cannabis* — 1
Robert C. Clarke

Introduction	1
Life Cycle	1
Origin, Early Evolution, and Domestication	5
Early History and Dispersal	7
Taxonomy	13
Brief History of Hemp Breeding	15
Conclusion	18

Chapter 2. The Phytochemistry of *Cannabis*: Its Ecological and Evolutionary Implications — 21
David W. Pate

Introduction	21
Cannabinoid Biogenesis and Anatomical Distribution	22
Cannabinoids and Environmental Stress	26
Evolution of Biogenetic Pathways	32
Conclusion	34

Chapter 3. Detecting and Monitoring Plant THC Content: Innovative and Conventional Methods — 43
Gianpaolo Grassi
Paolo Ranalli

Introduction	43
Immunodiagnostic	44
Polyclonal Antibodies	45
Monoclonal Antibodies	46

Recombinant Antibodies	46
Assay Format	47
Conclusion	58

Chapter 4. Agronomical and Physiological Advances in Hemp Crops — 61
Paolo Ranalli

Introduction	61
Cannabis Gene Pool	62
Variety Recommendations	63
Hemp in Crop Rotation	64
Seedbed	65
Methods of Planting	65
The Effect of Temperature on Leaf Appearance and Canopy Establishment in Fiber Hemp	66
Seeding Rate	67
Growing Conditions	69
Cultural Practices	69
Phenological Development	70
Effect of Nitrogen Fertilization and Row Width	72
The Chemical Composition of Hemp Stem	72
Cultivation Techniques and Crop Destination	74
Constraints to Dry Matter Production in Fiber Hemp	79
Implications for Future Research	80

Chapter 5. Crop Physiology of *Cannabis sativa* L.: A Simulation Study of Potential Yield of Hemp in Northwest Europe — 85
Hayo M. G. van der Werf
Els W. J. M. Mathijssen
Anton J. Haverkort

Introduction	85
Crop Physiological Characteristics	88
Potential Yield	99
Hemp versus Kenaf	102

Chapter 6. A Survey of Hemp Diseases and Pests   109
John M. McPartland

Introduction	109
Insect Pests of Stalk and Roots	110
Insect Pests of Leaves, Flowers, and Seeds	113
NonInsect Pests	115
Fungal Diseases	116
Other Diseases	121
Control of Diseases and Pests	122

Chapter 7. *Cannabis* Germplasm Resources   133
Etienne P. M. de Meijer

Introduction	133
The Structure of the *Cannabis* Gene Pool	134
Fiber and Seed Strains	141
Drug Strains	147
Other Domesticated *Cannabis*	148
Cannabis Germplasm *Ex-Situ*	148

Chapter 8. Genetic Improvement: Conventional Approaches   153
Ivan Bócsa

Historical Review	153
Sex Genetics	155
Improvement in Stem Yield	159
Breeding for an Increase in Fiber Content	164
Breeding for Reduced THC Content	175
Resistance Breeding	178
Breeding for Seed Yield and Oil Content	178
Biotechnological Aspects	179

Chapter 9. Advances in Biotechnological Approaches for Hemp Breeding and Industry   185
Giuseppe Mandolino
Paolo Ranalli

Tissue Culture and Breeding	186
Cell Culture and Secondary Metabolism	194
Molecular Markers for Hemp Breeding	197
Future Perspectives	208

Chapter 10. Alkaline Pulping of Fiber Hemp   **213**
Birgitte de Groot
Gerrit J. van Roekel Jr.
Jan E. G. van Dam

Introduction	213
Bast Fiber Pulps for Paper Applications	215
Hemp Woody Core and Hardwood Pulp	219
Introduction of Hardwood and Recycled Fibers in Paper	219
Important Pulping Processes and Their Significance for Hemp	222
Alkaline Pulping of Hemp	224
Recommendations	237

Chapter 11. Hemp Seed: A Valuable Food Source   **243**
David W. Pate

Introduction	243
Extraction Methods	245
Oil Composition and Properties	246
Critical Enzyme	248
GLA Importance	250
SDA Supporting Role	251
Future Prospects	252

Index   **257**

ABOUT THE EDITOR

Paolo Ranalli, PhD, graduated from the University of Bologna with a degree in Agricultural Science. He has been a researcher in the Experimental Institute for Industrial Crops (Bologna) since 1974. In 1980, he earned his PhD in plant breeding. In 1986, he passed a competitive exam to become Director of the Plant Breeding Section of the Experimental Institute for Industrial Crops; in 1997 he was appointed director of this institute.

Dr. Ranalli has taken part as project leader in many domestic and international projects engaged in classical and advanced breeding topics devoted to improve many agronomic traits of the plant (especially yield and resistance to biotic and abiotic stress).

He bred and released two varieties of peas for processing, one variety of pea for dry seed, seven varieties of common bean, and two varieties of potato. Most of them were included in the national list of varietal recommendation and are currently grown on a large scale. He also patented two new mutants in hemp for textile use.

He is the author or co-author of five textbooks and more than 200 papers on genetics, breeding, biotechnology, and crop physiology of grain legumes, potato, sugar beet, and hemp.

CONTRIBUTORS

Ivan Bócsa, PhD, DSc, is a scientific advisor at the Agricultural Research Institute Kompolt and Professor at the Godollo University for Agricultural Sciences, Hungary.

Robert C. Clarke, is the Projects Manager for the International Hemp Association, Amsterdam, Netherlands, works as a consultant to the hemp industry, and is the author of three books on *Cannabis*.

Birgitte de Groot, Ir, was a researcher on alkaline pulping technologies at Agrotechnological Research Institute (ATO-DLO), Wageningen, in the Dutch Hemp Project. He is currently process engineer for raw material quality at SCA-Molnlycke (Gennep) BV, Incontinence Care, Netherlands.

Gianpaolo Grassi, is a virologist and researcher at Experimental Institute for Industrial Crops, Bologna, Italy.

Anton J. Haverkort, PhD, is head of the crop science department of the Research Institute for Agrobiology and Soil Fertility (AB-DLO) in Wageningen, Netherlands.

Giuseppe Mandolino, is a biologist and researcher at the Experimental Institute for Industrial Crops, Bologna, Italy.

Els W. J. M. Mathijssen, is employed at DLO Research Institute for Agrobiology and Soil Fertility (AB-DLO) in Wageningen, Netherlands.

John M. McPartland, DO, MS, is a professor at the University of Vermont, and director of VAM/AMRITA, Middlebury, Vermont.

Etienne P. M. de Meijer, PhD, is a plant breeder at HortaPharm BV, Amsterdam, Netherlands.

David W. Pate, MS, is Senior Technical Officer at HortaPharm BV in Amsterdam and serves as Board Secretary for the International Hemp Association, and as editor of its journal. He is currently a doctoral student in the Pharmaceutical Chemistry Department at the University of Kuopo, Finland.

Jan E. G. van Dam, PhD, is senior researcher and industrial consultant on fiber crops processing and innovative product development in the department of agrofibers and cellulose at the Agrotechnological Research Institute (ATO-DLO), Wageningen, Netherlands.

Hayo M. G. van der Werf, PhD, is editor in chief of the *Journal of the International Hemp Association.* He is currently conducting a research program into the assessment of the sustainability of farming systems at the National Institute of Agronomic Research (INRA) at Rennes, France.

Gerrit J. van Roekel Jr., Ir, is research manager of the nonwood pulp and paper technology group at Agrotechnological Research Institute (ATO-DLO), Wageningen, Netherlands. He is an expert on mechanical pulping of annual fibers, and is a member of the TAPPI nonwood plant fiber committee.

Preface

The European community (EU), the United States, and other developed countries currently need to promote alternatives to crops produced in excess (such as cereals), and cultivations with a limited environmental impact. Hemp is able to satisfy these requirements, and can provide products (fiber and cellulose) in which the EU and other countries are deficient. Since there is no agreement about the admissibility of *Cannabis* cultivation among different countries, and—if cultivation is allowed—about tolerable cannabinoid levels, governmental and agricultural organizations can help to define regulations that will make it easier for farmers to grow this crop.

Hemp (*Cannabis sativa* L.) fiber can be used as a raw material for paper and textile production. It is a potentially profitable crop, having the right profile to fit into sustainable farming systems. Interest in "new" fiber crops is increasing, for example, to replace cotton with a less polluting alternative, or to relieve the pressure of the paper industry on remaining natural forests.

The oil content of hemp seed is high and compares in yields per hectare with rape and sunflower oil; it also has important pharmaceutical properties and finds a ready market in this application.

Since the renewed interest for this crop is increasing all over the world, it could be important to update the knowledge underlying this crop (i.e., whether it is capable of the productivity and fiber quality required and if not, what contribution plant breeding, agronomical practice, stress tolerance, and processing techniques are making, or could further make, to improve these characteristics). Thus, a review of development in the germplasm resources, genetics, and breeding objectives; breeding for resistance and quality; and improvement of physiological, morphological, and biochemical characteristics of hemp would be useful to promote research on this crop.

This book will provide interesting, convincing, and useful reading for all those who wish to work with hemp, as well as for those who teach about or study it. In particular, the book provides an overview of hemp and a reference guide on botany, phytochemistry, THC detecting methods, agronomical and physiological advances, diseases and pests, germplasm resources, genetic improvement by conventional and biotechnological approaches, pulp and paper production, and food provided by the seeds. This work is recommended to anyone who requires a basic knowledge of hemp cultivation and processing for textile and other production, especially students, scientists, growers, and vegetable breeders. If progress is to be made in breeding for enhanced yield, disease resistance, and other traits, it will be achieved through an integration of conventional practices, a better understanding of constraints and how to overcome them, and the application of molecular approaches. Although all the chapters were prepared by specialists in the field, the differences in approach sometimes provide a striking contrast among more local versus regional or international perspectives.

By bringing together information on the different topics, their problems, and the approaches being taken for their improvement, we hope to provide scientists, teachers, students, and extension workers with a core of information that can be used in the improvement of hemp.

Paolo Ranalli
Istituto Sperimentale per le Colture Industriali
Via di Corticella 133, I-40129 Bologna (Italy)
Phone: +39-51-6316847
Fax: +39-51-6316847
e-mail: ranalli@bo.nettuno.it

Chapter 1

Botany of the Genus *Cannabis*

Robert C. Clarke

INTRODUCTION

This chapter explores the life cycle of *Cannabis*, its origin, early evolution and domestication, the dispersal and taxonomy of *Cannabis* landraces and a brief history of *Cannabis* breeding. *Cannabis* is among the very oldest of economic plants, and humans have long been attracted to its multiple uses. Many landraces have evolved resulting from varying human selective pressures on *Cannabis* as a provider of hemp fiber, edible seed, and resins for drug use. Human selection has operated in concert with natural selective pressures imposed by the diverse environments into which humans have introduced *Cannabis*. Much more recently, innovative classical breeding techniques have been used to improve fiber or seed *Cannabis*, resulting in many high-yielding fiber cultivars suitable for temperate climates. No hemp fiber or seed cultivars exist for subtropical or tropical regions. In the future, we may see the use of advanced molecular techniques combined with innovative breeding strategies to develop *Cannabis* cultivars for many uses.

LIFE CYCLE

Whether *Cannabis* grows wild or is cultivated for its fiber, seed, or drug, its natural life cycle is the same. *Cannabis* is a medium to tall, erect, annual herb. However, environmental influences on the growth habit of *Cannabis* are very strong. Provided with an open sunny environment, light well-drained soil, and ample nutrients and water,

Cannabis can grow to a height of 5 m in a four- to six-month growing season. Exposed riverbanks, meadows, and agricultural lands are ideal habitats for *Cannabis* since all offer good sunlight. When growing in arid locations with limited soil nutrients, *Cannabis* plants develop minimal foliage and may mature and bear seed when only 20 cm tall. When planted in close stands, as for fiber hemp cultivation, *Cannabis* plants do not branch but grow as tall, thin, straight stalks. If an individual plant is not crowded by its neighbors, as in seed production, limbs bearing flowers will grow from small buds located at the nodes (intersection of the petioles or leaf stalks) along the main stalk.

Illustration 1.1 demonstrates that both of these plants are of the "Novosadska Konoplya" variety and are grown under the same field conditions. The plant on the left was grown at a field density of about four plants/m^2 for seed production and the plant on the right was grown at a density of about 100 plants/m^2 for fiber production.

Seeds are sown outdoors in the spring and usually germinate in three to seven days. About 10 cm or less above the cotyledons (seed leaves), the first true leaves arise, a pair of oppositely oriented single leaflets. Subsequent pairs of leaves arise in opposing pairs and a variously shaped leaf sequence develops with the second pair of leaves having three leaflets, the third five, and so on up to eleven to thirteen leaflets. Under favorable conditions *Cannabis* can grow up to 10 cm a day in height during the long days of summer.

Cannabis exhibits a dual response to daylength. During the first two or three months of growth it responds to increasing daylength with more vigorous vegetative growth, but later in the same season *Cannabis* requires shorter days to flower and complete its life cycle. *Cannabis* flowers when exposed to a critical daylength of twelve to fourteen hours, which varies with the strain, depending on its latitude of origin. Most strains have an absolute requirement for a minimum number of inductive daylengths (short days or more accurately long nights) that will induce fertile flowering. Fewer inductive daylengths than this will result in the formation of undifferentiated primordia (unformed flowers) only. Dark (night) cycles must be uninterrupted by light periods in order to induce flowering.

Cannabis is normally a dioecious plant, which means that the male and female flowers develop on separate plants, although monoecious examples with flowers of both sexes on one plant are occasionally

Illustration 1.1

Source: Courtesy of Dr. Janos Berenji.

found. The development of branches bearing flowering organs varies greatly between males and females. The male flowers hang in long, loose, multibranched, clustered panicles up to 30 cm long, while the female flowers are tightly crowded in the axils (junctions of small leaves and the central stem) within the erect racemes. *Cannabis* is anemophilous (wind pollinated) and relies on air currents to carry the pollen grains from the male plants to the female plants.

The first sign of flowering in *Cannabis* is the appearance of single undifferentiated flower primordia along the main stalk at the nodes, behind each of the stipules (leaf spurs). Before flowering, the sexes of *Cannabis* are indistinguishable, except for general trends in growth habit in certain strains such as height and extent of branching. After flowering is induced the male flower primordia can be identified by their curved claw shape, soon followed by the differentiation of round, pointed flower buds having five radial segments. The female primordia are recognized by the enlargement of a symmetrical tubular bract (floral sheath).

In uncrowded conditions the female plants tend to be shorter and have more branches than the male. Female plants are leafy to the top with many small leaves surrounding the flowers, while male plants have fewer leaves near the top with very few small leaflets along the elongated flowering limbs. The female flowers appear as two long white, yellowish, or pinkish stigmas protruding from the fold of a very thin membranous bract. The bract is covered with resin-exuding glandular trichomes (hairs). Female flowers are borne in pairs at the nodes on each side of the petiole behind the stipule which conceals the small flower. The bract measures 2 to 8 mm in length and is closely applied to, and completely contains, the ovary.

In male flowers, five petals approximately 5 mm long make up the corolla and are usually light yellow to greenish in color. They hang down, and five stamens approximately 5 mm long emerge, consisting of slender anthers (pollen sacs), splitting upward from the tip and suspended on long filaments. The pollen grains are nearly spherical, slightly yellow, and 25 to 30 µm in diameter. The surface is smooth and exhibits two to four germ pores.

In both sexes, before the start of flowering the phyllotaxy (leaf arrangement) reverses from opposite to alternate and usually remains alternate throughout the floral stages regardless of sexual type. Also the number of leaflets per leaf decreases until a small single leaflet appears below each pair of flowers.

The differences in flowering patterns of male and female plants are expressed in many ways. Soon after pollen is shed, the male plant usually dies. The female plant may mature up to five months after viable flowers are formed, if little or no fertilization occurs, and it is not killed by frost. Compared with female plants, male plants show a more rapid increase in height and a more rapid decrease in leaf size to the leaflets that accompany the flowers.

Many factors contribute to determining the sexuality of a flowering *Cannabis* plant. Under average conditions, with a normal inductive daylength, *Cannabis* populations will flower and produce approximately equal numbers of male and female plants, their sex determined by simple X and Y sexual inheritance. Monoecism (male and female flowers on the same plant) is an aberration that has been used by breeders to create relatively stable monoecious hemp cultivars. However, under modifying conditions of extreme stress, such as nutrient excess or deficiency, mutilation, extreme cold, or radically altered light cycles, populations have been shown to depart greatly from the expected one-to-one, male-to-female sex ratio, and monoecious individuals of many different phenotypes may arise.

Pollination of the female flower results in the loss of the paired stigmas and a swelling of the tubular bract where the ovule is enlarging. After approximately three to six weeks the seed is matured and after some time is harvested and dispersed by humans or drops to the ground. This completes the normal four- to six-month life cycle, which may take as little as two months or as long as ten months. Fresh fully mature seeds approach 100 percent viability, but this decreases quickly with age. Usually at least 50 percent of the seeds will germinate after three to five years of storage at room temperature.

The mature achene fruit (commonly referred to as a seed) is partially surrounded by the bract. The perianth (seed coat) is variously patterned in gray, brown, or black. Elongated and slightly compressed, the seed measures 2 to 6 mm in maximum length and 1 to 4 mm in maximum diameter.

Illustration 1.2 demonstrates three seed accessions ranging from a thousand seed weight of 3 g for the tiny wild seeds from Shandong Province, China, up to 60 g for the very large "snack food" seed landrace from Yunnan Province, China. The intermediate-sized seeds are "Kompolti Hibrid TC" from Hungary with a thousand seed weight of about 20 g.

ORIGIN, EARLY EVOLUTION, AND DOMESTICATION

The exact origin of *Cannabis* is unclear, because it was dispersed across Eurasia by humans very early in pre-history. However, Central

6 ADVANCES IN HEMP RESEARCH

Illustration 1.2

Asia offers by far the most plausible location for the origin of *Cannabis*. From Central Asia *Cannabis* was carried throughout East Asia, South Asia, and Europe which served as primary centers of domestication and secondary gene pools.

Humans have long been attracted to the many economically valuable characteristics of *Cannabis*. Fiber, food and oil seeds, and drugs have been the most important primary plant products derived from *Cannabis*. It is impossible to say with certainty which of these products were used first. Ancient humans must have discovered early the virtues of the *Cannabis* plant. Its vigorous growth makes it stand out and its unique appearance makes it readily distinguishable. The economically valuable attributes of *Cannabis* are obvious. Its fibers persist on the ground after the stem rots, the fruits are relatively large and prevalent, and the resin glands sparkle in the sun and adhere to the skin when handled. The human roles in initially selecting wild *Cannabis*, developing cultivars for varying uses, and insulating them from wild populations through continued cultivation, isolation, and selection, have been largely responsible for the present pattern of diversity in *Cannabis*.

The use of *Cannabis* varies between cultures. European, northern Asian, and eastern Asian cultures focused on the potential of *Cannabis* as a fiber and edible seed producer. African, Middle Eastern, South Asian, and Southeast Asian cultures used *Cannabis* primarily as the source of a psychoactive drug and secondarily as a fiber and seed plant. Fiber landraces originated in Europe and East Asia. Drug landraces originated in South Asia and spread into Southeast Asia and South and East Africa in approximately 100 A.D. All of these regions have high-THC (Δ^9-tetrahydrocannabinol) landraces nearly devoid of nonpsychoactive cannabidiol (CBD). This likely results from continued selection for psychoactive potency after dispersal to the new regions. Natural selection for high THC content in regions with high ultraviolet (UV) light levels may also have contributed to the evolution of high-THC landraces (Pate, 1994).

Western Europe and East Asia have both served as origins for the dispersal of hemp fiber varieties to the New World. The initial dispersals were from Western Europe during the 1600s. Later dispersals were from China and Japan during the early 1900s. Intensive breeding of fiber varieties continued in the United States from the late 1800s into the 1930s. Presently, hemp variety development in Europe has resumed its progress as hemp cultivation increases worldwide.

Over the past several thousand years, trade and cultural isolation have had varying effects on mixing and separating sections of the *Cannabis* gene pool. Dispersal of drug varieties in a northerly direction from India was precluded by northern cultural preferences and local requirements for hemp fiber and seed, resulting in cultural, and therefore geographical, constraints on THC levels. Likewise, fiber varieties from the north were rarely taken to more equatorial latitudes, as native fiber crops were abundant in these regions and consequently hemp fiber was not of much value. Cultural preferences and resulting human selections have had the greatest effect in modifying and preserving landrace and variety characteristics.

EARLY HISTORY AND DISPERSAL

Cannabis was among the earliest cultivated plants and for centuries ranked as one of the most important agricultural crops. Prior to

1000 B.C. until the late 1800s, *Cannabis* was used to produce myriad necessities such as cordage, cloth, food, lighting oil, and medicine and was one of the most widely cultivated plants. *Cannabis* hemp is the strongest and most durable of natural fibers and is versatile enough to be used for the manufacture of fine cloth as well as rope, twine, canvas, and sacking. Until the popularization of the cotton gin, hemp was the most widely used textile fiber. The invention of petrochemical fibers and acid-process pulp paper lessened the importance of *Cannabis* hemp and drastically lowered its ranking as a world crop. Kerosene replaced *Cannabis* seed as a source of lighting oil. *Cannabis* has been recommended in the treatment of diverse medical conditions for over 3,000 years. *Cannabis* extracts were the analgesics most widely prescribed by Western doctors during the 1800s and early 1900s, and *Cannabis* offers promise as a source of future medicines. The prohibition of *Cannabis* drugs has led to the prohibition of *Cannabis* cultivation in general, and the historically important uses of *Cannabis* have been largely forgotten, lost behind a smoke screen of unresolved controversy.

Historical information gives us a good picture of *Cannabis'* dispersal from Central Asia into Europe, Africa, and finally the New World. China has produced the oldest archeological and historical evidence for the antiquity of *Cannabis*. It is likely that the Chinese were the first to use wild *Cannabis* and domesticate it for its fiber and seed, and that the Indians were the first to use it and domesticate it for its psychoactive properties (Abel, 1980).

Illustration 1.3 demonstrates that the primary dispersal of *Cannabis* was from its origin in Central Asia into East Asia, South Asia, and Europe. Much later on, many secondary and tertiary dispersals of *Cannabis* germplasm carried fiber landraces from East Asia and Europe, and drug landraces from South Asia, throughout Africa and the New World.

Early in their history, the Chinese developed wild *Cannabis sativa* into fiber- and seed-producing landraces and *Cannabis* made a major contribution to northern China's early agricultural economy. Shortly thereafter the South Asian peoples discovered *Cannabis*, and with the help of a favorable climate, selected and developed *Cannabis* for its psychoactive potential. Prior to the Roman era, Europeans finally

Illustration 1.3a

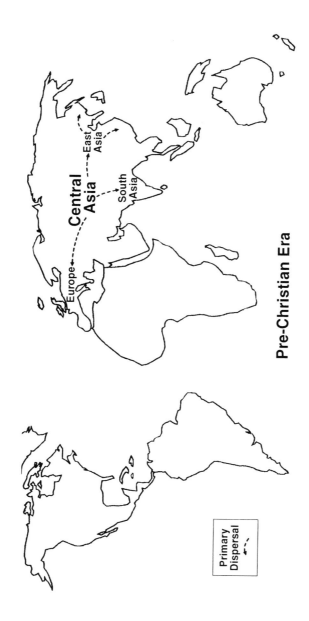

Illustration 1.3b

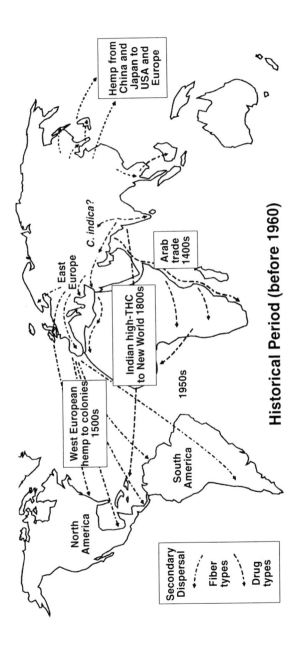

Illustration 1.3c

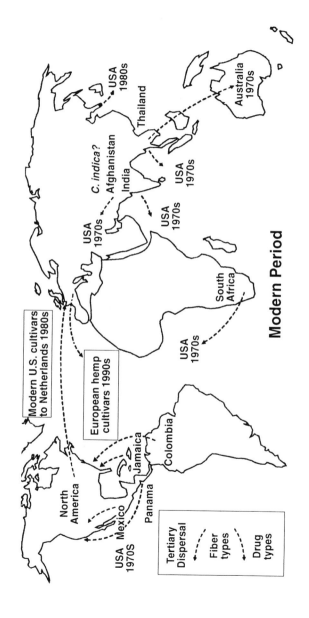

began to use wild *Cannabis* as a fiber and seed plant. All three major secondary gene pool sections—Indian drug-type *Cannabis*, Chinese fiber-type *Cannabis*, and European fiber-type *Cannabis*—apparently evolved before the time of Christ. These three initially derived secondary gene pools were the progenitors of modern cultivated *Cannabis* varieties.

Fiber landraces of *Cannabis* containing little of the psychoactive constituent THC, were initially brought to North America as early as 1545 for rope and sail manufacture by British, French, and Spanish colonists (Abel, 1980) where they escaped cultivation and became weeds in the northern midwest states and southern central Canada. Northern European hemp was adapted to a relatively narrow range of climatic conditions, and although its fiber yield was relatively low and it was apparently better suited for seed production, it was the only hemp variety grown in America until the early 1890s. During the late 1800s and early 1900s, Japanese hemp was introduced into California and Chinese hemp into Kentucky. Both these introductions were Asian *Cannabis* fiber landraces and represented independently evolved sections of the *Cannabis* gene pool distinct from European fiber *Cannabis*. Chinese hemp landraces were acclaimed from the start as superior. The Chinese perfected the weaving of hemp cloth early in their history and the Japanese still produce hemp cloth nearly as sheer as fine silk. The Chinese introductions into Kentucky became known as the 'Kentucky' hemp variety.

The South Asian section of the *Cannabis* gene pool was improved through selection as a psychoactive drug plant. By the first century A.D., Indian high-THC varieties had been spread by traders from the tip of southern Africa to Sumatra and then they slowly spread inland across sub-Saharan Africa. In 1835, *Cannabis* began its diffusion to the Caribbean and northern South America from India. Indian indentured servants and laborers are thought to have introduced high-THC landraces to the New World. Indian *ganja* (marijuana) landraces spread into Central America from the Caribbean. Eventually high-THC drug varieties spread to the United States during the 1960s and 1970s in a steady stream of shipments of illicit marijuana from Jamaica, Mexico, Colombia, and Southeast Asia. Many landraces, and especially those from Afghanistan and Pakistan, have contributed to the modern drug type *Cannabis* cultivars.

TAXONOMY

Cannabis and *Humulus* (and possibly *Humulopsis*) (Grudzinskaya 1988) are the only genera in the family Cannabaceae. *Cannabis* either grows spontaneously or is cultivated throughout nearly all equatorial to subarctic regions of the world. According to Zhukovskii (1964), *C. sativa* grows wild in river basins and on slopes in the Transvolga and islands of the Volga Delta as well as in the Himalaya, Hindu Kush, Tien Shan, and Altai mountains. Despite heated debate over the species status of *Cannabis*, the question of species assignment has never been approached from a modern broad perspective and remains unclear. In addition, semantics and legal issues have played significant roles in confusing species disputes. The thorough characterization of extant *Cannabis* taxa and interpretation of their evolutionary interrelationships has yet to be achieved.

Cannabis has been variously characterized by modern taxonomists. Some favor lumping all *Cannabis* taxa together into a single species *C. sativa* circumscribing two subspecies, each further divided into two varieties (Small and Cronquist, 1976). Others favor splitting *Cannabis* into three species: *C. sativa*, *C. indica*, and *C. ruderalis*, each species circumscribing its own varieties (Schultes et al., 1974). Still others do not recognize *C. ruderalis* but preserve the first two species (Vavilov and Bukinich, 1929; Serebriakova, 1940; Zhukovskii, 1950). In all three systems of nomenclature, *C. sativa* is the most varied and geographically diverse taxon and circumscribes the majority of fiber, seed, and drug varieties.

C. sativa L. is usually characterized by its tall stature, less developed branching especially in the fiber varieties, light to medium green foliage, and a pronounced spicy or sweet aroma. Drug varieties secrete resins of high cannabinoid content predominating in THC while fiber varieties secrete resin which is much lower in total cannabinoids and particularly low in THC. The vast majority of the spontaneous populations of *Cannabis* have also been classified as members of species *sativa*.

The type specimen of *C. indica* chosen by Lamarck is a narrow-leaved variety with a relatively elongated loose inflorescence. The epithet *"indica"* literally means "from India" and this is apparently what Lamarck intended when he named the species. Around the turn of the century the name *Cannabis indica* came to represent the pharmaceutical *Cannabis* imported from India. This followed Lamarck's intentions exactly as pharmaceutical *Cannabis* originated in India. However, Lamarck's type specimen is visually indistinguishable from drug strains of *C. sativa*.

Vavilov and Buckinich (1929) described two broad-leaved varieties from Afghanistan and Pakistan and included them as members of *C. indica* Lam. At the time, this must have seemed correct, as these varieties differed greatly from European hemp varieties of *C. sativa*, therefore requiring, in Vavilov's mind, that they be assigned to a separate taxa, and they were native to the Indian subcontinent. Vavilov did not mention if the Afghan varieties were drug or fiber types. Zhukovskii (1964) assigns *C. indica* a wild range of Pakistan and Afghanistan and a cultivated range of India, Iran, Turkey, Syria, and North Africa.

Schultes also accepts a broad interpretation of *C. indica* and includes Afghan as well as Indian drug types and their New World descendants within *C. indica* Lam. The Afghan *Cannabis*, described by both Vavilov and Bukinich (1929) and by Schultes et al. (1974), a short, broad-leaved, and acrid-smelling drug plant, has come to typify *C. indica*, especially in the eyes of marijuana growers who use the term "indica" to differentiate Afghan drug varieties from all other drug varieties or "sativas" (Clarke, 1981).

The Indian drug types described by Lamarck and included in *C. indica* by some subsequent researchers (Schultes et al., 1974) are actually drug varieties of *C. sativa* indigenous to South Asia (Small and Cronquist, 1976). In fact, the varieties Vavilov and Buckinich (1929) and Schultes et al. (1974) described from Afghanistan should not be properly included in *C. indica* either, since they were not from India and were not at all similar to Lamarck's type specimen. The evolutionary history and taxonomic classification of Afghan *Cannabis* has yet to be accurately determined.

Some researchers have hypothesized that *C. ruderalis* is a truly wild taxon and was the ancestor to cultivated varieties of *Cannabis* (Vavilov, 1931; Zhukovskii, 1964), but it could also be a feral variety escaped from cultivation. It may also be classified as *C. sativa* ssp. spontanea as described by Serebriakova (1940).

BRIEF HISTORY OF HEMP BREEDING

Throughout the course of history, humans have rarely been satisfied with the qualities of wild *Cannabis* as a source of fiber, seed, and/or drug. Wild populations were most likely depleted near settlements by persistent annual collection and ancient humans may have found it more convenient to cultivate *Cannabis* near home, rather than travel farther and farther each year in search of wild harvests. Early farmers realized the superiority of cultivated plants and soon favored semidomesticated landraces over their wild relatives.

Only a very limited number of farmers, regardless of whether they grow *Cannabis* for its fiber, seed, or drug, consciously select and breed their plants in an effort to improve them. European hemp breeders continue to develop improved hemp fiber cultivars and clandestine marijuana breeders secretly work to improve drug types of *Cannabis*, but the vast majority of *Cannabis* growers worldwide practice no selection at all.

Cannabis is not a particularly straightforward plant to breed. The life history of *Cannabis* presents two major obstacles to its improvement by selective breeding. *Cannabis* is usually dioecious, and thus *Cannabis* plants are usually incapable of pollinating themselves, and are therefore outcrossing by nature. Selfing or inbreeding is the most effective means of fixing desirable traits, since the selected genes are more likely to be represented in both the pollen and the ovule if they come from the same plant. However, in obligate outcrossers such as *Cannabis*, persistent inbreeding is practically difficult and results in a loss of vigor. In traditional dioecious *Cannabis* breeding, the genes controlling a selected trait must be present in two separate individuals, one male pollen parent and one female seed parent. The economically valuable products of *Cannabis*, whether fiber, seed, or drugs, all come primarily from female plants. This makes it very difficult to recognize potentially favorable traits in male parents and these traits must then ultimately be expressed in the female offspring. Also, since all *Cannabis* is anemophilous (wind-pollinated) and intercrosses freely, prospective seed parents must be isolated to avoid stray pollinations until they are to be pollinated with a selected male. This requires isolation by geographical distance or mechanical means.

Because *Cannabis* is a difficult plant in which to fix traits through selective breeding, and often only the female plants are of economic importance, modern marijuana growers have found that it is advantageous to clone exceptional plants by rooting stem cuttings. In this way practically unlimited numbers of identical select plants can be grown. Besides circumventing the vagaries of genetic recombination, cloning can produce uniform crops of female plants in one generation. Clones of valuable male plants can also be preserved for pollen production in breeding programs. Vegetative clone libraries are maintained under artificially created long daylength.

European hemp breeders have developed many highly productive fiber cultivars with low THC content. Breeders in France, Germany, Poland, Romania, and the Ukraine have concentrated primarily on developing even-maturing, low-THC monoecious varieties, while those in Hungary, Italy, Spain, and Yugoslavia have concentrated primarily on low-THC dioecious cultivar development. Some of the most innovative cultivars were developed by Ivan Bócsa at the G.A.T.E. Rudolf Fleischmann Research Institute in Kompolt, Hungary. Bócsa applied a method developed by Bredemann in Germany for the selection of male plants for fiber content prior to flowering, resulting in large increases in fiber percentage, and was the first to develop high-yielding hybrids and unisexual female generations.

The forty-five currently registered European fiber hemp cultivars were developed by starting with a few promising local landraces unconsciously selected over hundreds of years by peasant farmers in areas of traditional hemp cultivation. The "modern" industrial hemp varieties are descendant from very few ancestors. The European varieties are based on three gene pool sections: Northern and Central European ecotypes, Southern European ecotypes, and East Asian ecotypes (de Meijer, 1995). The thirty-two commercially available registered hemp varieties consist of twenty-two monoecious, nine dioecious, (and one unisex female) sexual types. All thirty-two originated entirely or in part from landraces of the Central and Southern European ecotypes and only two of the Hungarian varieties incorporate ancestors from the Far Eastern ecotype. All of the monoecious and unisex varieties derived their monoecious trait from 'Fibrimon.' Seven of the nine dioecious varieties include genes from Hungarian landraces and other varieties that were all originally derived from

Italian landraces. A single Far Eastern accession from China was used to establish the hybrid triple cross varieties from Hungary. Italian landraces combined with the single monoecious line 'Fibrimon', and a single Chinese landrace along with a few other Central and Southern European accessions were used to breed all of the industrial hemp varieties. This is a narrow genetic base and possibly explains why industrial hemp varieties are so poorly suited to growing in other regions of the world beyond Europe.

Until recently there has been little incentive to develop new *Cannabis* varieties. Hemp was used almost entirely for fiber production and there were few commercial uses for the seed. During the 1980s, hemp cultivation nearly ceased in eastern Europe and there was little need for sowing seed of existing cultivars, much less the costly development of new ones. The resurgence of interest in hemp is accompanied by a wide range of new products and potential uses for hemp. Now, there is sufficient economic interest in hemp to warrant the development of special varieties with higher quality fiber, increased seed and seed oil yield, modified seed oil fatty acid profiles, specific cannabinoid profiles, suitability for equatorial cultivation, resistance to specific pests and diseases, salt tolerance, etc.

Hemp breeders have recently developed modern oilseed and pharmaceutical varieties for specific end uses and several new varieties are entering production. 'FIN-314', the first hemp cultivar strictly for grain seed production, was developed in Finland from germplasm accessioned in the Vavilov Research Institute Gene Bank and entered commercial production in Canada in 1998. The seed oil of 'FIN-314' contains high levels of the essential fatty acids *gamma*-linolenic acid (4.4 percent) and stearidonic acid (1.7 percent), which are valuable as nutraceuticals and dietary supplements (Laakkonen and Callaway, 1998). The short stature of 'FIN-314' makes it readily harvestable by machine combine and seed yields are expected to be high. Further improvements may include selection for consistent early seed set and determinant crop ripening, which could increase seed yield and decrease the percentage of immature seeds.

Pharmaceutical *Cannabis* varieties have been developed in the Netherlands by HortaPharm BV and have been licensed to GW

Pharmaceuticals Ltd. for production in England. These cultivars will be used to prepare a variety of pure cannabinoid extracts and whole plant tinctures that will be tested in clinical trials for medical efficacy for several indications. The varieties developed by Horta-Pharm produce single cannabinoids such as CBD and THC at very high levels (over 10 percent) without significant amounts of any other cannabinoids. This will allow the testing of crude *Cannabis* with a reproducible cannabinoid profile as well as simple formulations of blended cannabinoid profiles. Also, the production of THC from natural sources should be much less expensive than the current method of production by laboratory synthesis.

CONCLUSION

Cannabis originated in Central Asia and was dispersed by humans throughout the world from subarctic to tropical regions. The strong fibers, edible fruits, and psychoactive drugs produced by *Cannabis* have attracted humans since Neolithic times. Although *Cannabis* has been cultivated for several thousand years, it remains incompletely domesticated, and weedy escapees from cultivation are relatively common.

Historical data support the contention that *Cannabis* was spread by humans from Central Asia into the remainder of Eurasia. Wild ancestral populations can still be found in Central Asia.

Domestication has had the greatest influence on the evolution of *Cannabis*. Disruptive human selection and isolation of varieties in cultivation have contributed to the diversity experienced in modern *Cannabis*. Humans have selected varieties specifically for fiber and/or seed uses or drug use and each group of strains is on average morphologically and physiologically distinct from the others.

Human selection has affected seed yield, fiber yield and quality, and cannabinoid profile. THC levels have primarily been enhanced by human selection for potency, much more so than by natural selective pressures, such as ultraviolet light levels.

Modern hemp fiber varieties all come from northern Europe and are derived from a very narrow genetic base. Breeders are currently developing advanced cultivars intended for new uses and climatic requirements.

Since lawmakers continue to confuse fiber varieties of *Cannabis* with drug varieties, hemp will likely not be grown again on a large scale in the United States for some time, even though it is increasingly cultivated throughout Canada and Eurasia. It is unreasonable to fear social repercussions from the rampant smoking of hemp, since it nearly always contains an insufficient amount of THC to cause any psychological effect (less than 1 percent d.w. THC). It is unfortunate that the future cultivation of industrial hemp is restricted by laws intended to control marijuana cultivation.

REFERENCES

Abel, E. (1980). *Marihuana: The first 12,000 years.* New York: Plenum Press, 100.
Clarke, R. C. (1981). *Marijuana botany.* Berkeley, CA: And/Or Press.
Grudzinskaya, I. A. (1988). "On the taxonomy of the Cannabaceae" [Russ.] *Journal of Botany*, 73(4):589-593.
Laakkonen, T. T. and Callaway, J. C. (1998). "Update on FIN-314." *Journal of the International Hemp Association*, 5(1):34-35.
Meijer de, E. (1995). "Fibre hemp cultivars: A survey of origin, ancestry, availability and brief agronomic characteristics." *Journal of the International Hemp Association*, 2(2):66-73.
Pate, David W. (1994). "Chemical ecology of *Cannabis*." *Journal of the International Hemp Association*, 1(2):34-35.
Schultes, R. E., Klein, W. M., Plowman, T., and Lockwood, T. E. (1974). "*Cannabis*: An example of taxonomic neglect." Botanical Museum Leaflets 23(9): 337-364, Harvard University.
Serebriakova, T. I. (1940). "V. Fiber Plants Part 1." In Wulff, E. V. (Ed.), *Flora of cultivated plants.* Moscow and Leningrad: State Printing Office.
Small, E. and Cronquist, A. (1976). "A practical and natural taxonomy for *Cannabis*." *Taxonomy*, 25(4):405-435.
Vavilov, N. I. (1931). "The role of Central Asia in the origin of cultivated plants"[Russ.], *The Bulletin of Applied Botany of Genetics and Plant Breeding*, 26:42.
Vavilov, N. I. and Bukinich, D. D. (1929). "Agricultural Afghanistan" [Russ.], *The Bulletin of Applied Botany of Genetics and Plant Breeding*, Supplement #33:378-382, 474, 480, 584-585, 604.
Zhukovskii, P. M. (1964). *Cultivated plants and their wild relatives*, Third edition, pp. 421-422. Leningrad: Kolos.

Chapter 2

The Phytochemistry of *Cannabis:* Its Ecological and Evolutionary Implications

David W. Pate

INTRODUCTION

Cannabis may have been the first cultivated plant. Records indicate use of this plant for paper, textiles, food, and medicine throughout human history (Abel, 1980). It is a dioecious annual with rather distinctive palmate leaves, usually composed of an odd number of leaflets. Best growth occurs on recently disturbed sites of high soil nitrogen content, so it is commonly found as a persistent weed at the edge of cultivated fields. Mature height ranges from 1 to 5 m, according to environmental and hereditary dictates. Typically, the male plant is somewhat taller and more obviously flowered. These flowers have five yellowish tepals and five anthers that hang pendulously at maturity, dispersing their pollen to the wind. The female plant exhibits a more robust appearance due to its shorter branches and dense growth of leaves and flower-associated bracts. The double-styled female flower possesses only a thin, closely adherent perianth, but is further protected by enclosure in a cuplike bracteole (i.e., perigonal bract), subtended by a usually monophyllous leaflet. The single achene produced per flower is shed or dispersed as a result of bird predation. The life cycle of the male is completed soon after anthesis, but the female survives until full seed ripeness.

Many thanks are due to Raphael Mechoulam and Yukihiro Shoyama for their gracious critiques of the manuscript.

Cannabis seems a virtual factory for the production of secondary metabolic compounds (Turner, ElSohly, and Boeren, 1980). A variety of alkanes have been identified (Adams and Jones, 1973; De Zeeuw, Wijsbek, and Malingre, 1973; Mobarak, Bieniek, and Korte, 1974a, 1974b), as well as nitrogenous compounds (ElSohly and Turner, 1976; Hanus, 1975a; Mechoulam, 1988), flavonoids (Gellert et al., 1974; Paris, Henri, and Henri, 1975; Paris and Paris, 1973) and other miscellaneous compounds (Hanus, 1976a, 1976b). Terpenes appear in abundance (Hanus, 1975b; Hendricks et al., 1975) and contribute to the characteristic odor of the plant (Hood, Dames, and Barry, 1973) and some of its crude preparations such as hashish. The compounds which comprise the active drug ingredients are apparently unique to this genus and are termed cannabinoids. Cannabinoids were originally thought to exist as the phenolic compounds, but later research (Fetterman, Doorenbos et al., 1971; Masoud and Doorenbos, 1973; Small and Beckstead, 1973; Turner et al., 1973) indicated their existence predominantly in the form of carboxylic acids which decarboxylate readily with time (Masoud and Doorenbos, 1973; Turner et al., 1973) upon heating (De Zeeuw, Malingre, and Merkus, 1972; Kimura and Okamoto, 1970) or in alkaline conditions (Grlic and Andrec 1961; Masoud and Doorenbos, 1973). Over sixty of these type compounds have been found in the genus (Turner, ElSohly, and Boeren, 1980).

Much has been published concerning the influence of heredity on cannabinoid production (Fetterman, Keith et al., 1971; Small and Beckstead, 1973), but ecological factors have long been thought to have an important influence by stressing the *Cannabis* plant (Bouquet, 1950). The resultant increased biosynthesis of the cannabinoid and terpene containing resin, in most cases, seems likely of advantage to the organism in adapting it to a variety of survival-threatening situations. This work reviews these biotic and abiotic challenges and speculates on the utility of *Cannabis* resin to the plant.

CANNABINOID BIOGENESIS AND ANATOMICAL DISTRIBUTION

The cannabinoid *delta*-9-tetrahydrocannabinol (THC) is responsible for the main psychoactive effects of most *Cannabis* drug preparations (Mechoulam, 1970). In some varieties of *Cannabis*, additional cannabinoid homologs appear that have the usual pentyl group attached to the aromatic ring, replaced by a propyl (De Zeeuw et al., 1972; De Zeeuw et al., 1973; Fetterman and Turner, 1972; Gill, 1971; Gill, Paton, and Pertwee, 1970; Merkus, 1971; Vree et al., 1972a; Turner, Hadley, and Fetterman, 1973) or occasionally a methyl (Vree et al., 1971, 1972b) group. Other claims have been made for butyl (Harvey, 1976) or heptyl (Isbell, 1973) substitutions, but seem tenuous, particularly in the latter case. THC was long thought (e.g., Mechoulam, 1970) to be produced by the plant (see Figure 2.1) from cannabidiol (CBD), which, in turn, is derived from cannabigerol (CBG) generated from noncannabinoid precursors (Fellermeier and Zenk, 1998; Turner and Mahlberg, 1988). This hypothetical scheme has also been tentatively supported by experimental evidence (e.g., Shoyama, Hirano, and Nishioka, 1984). However, recent work by Taura, Morimoto, and Shoyama (1995) has demonstrated CBG to be the direct biogenetic precursor of THC, at least in some strains. CBG is also the biogenetic

Figure 2.1. Proposed biogenesis of the major cannabinoids.

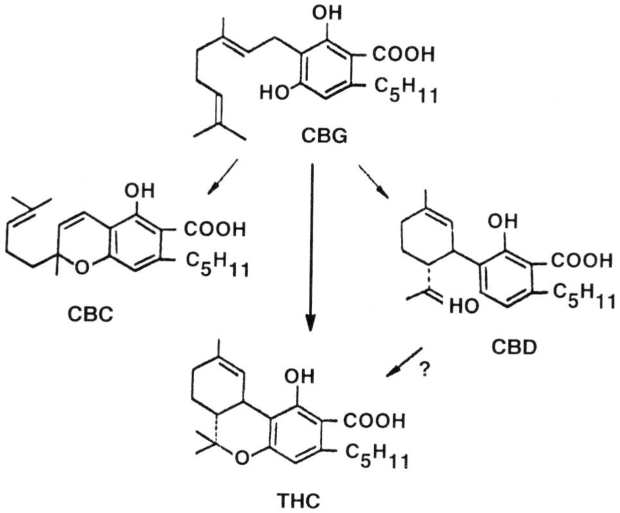

Source: Adapted from Shoyama, Y., Yagi, M., Nishioka, I., and Yamauchi, T. (1975). "Biosynthesis of cannabinoid acids." *Phytochemistry*, 14:2189-2192.

precursor of cannabichromene (CBC). Some of the cannabinoids (e.g., cannabielsoin, cannabinol, and cannabicyclol) are probably physical conversion products of the enzymatically produced cannabinoids (e.g., CBD, THC, and CBC, respectively).

The major sites of cannabinoid production appear to be epidermal glands (Fairbairn, 1972; Hammond and Mahlberg, 1973; Lanyon, Turner, and Mahlberg, 1981; Malingre et al., 1975) which exhibit a marked variation in size, shape, and population density, depending on the anatomical locale examined. While there are no published reports of glands present on root surfaces, most of the aerial parts possess them, along with nonglandular trichomes (De Pasquale, Turmino, and De Pasquale, 1974). These epidermal glands seem to fall into two broad categories: stalked and sessile. The stalked gland (see Figure 2.2) can consist of a single cell or small group of cells arranged in a rosette on a single or multicellular pedestal. Lack of thorough ontogenetic study has led to the speculation that some of this variation may be attributable to observation of various developmental stages (Ledbetter and Krikorian, 1975). The sessile gland possesses no stalk and has secretory cells located at or below the epidermal surface (Fairbairn, 1972). In either case, the glandular cells are covered with a "sheath" under which the resins are secreted via vesicles (Mahlberg and Kim, 1992). This sheath consists of a cuticle that coats a polysaccharide layer (presumed cellulose) originating from the primary cell wall (Hammond and Mahlberg, 1978). The resins accumulate until the sheath bulges away from the secretory cells, forming a spheroid structure. The resin is then released via rupture of the membrane or through pores in its surface (De Pasquale, Tumino, and De Pasquale, 1974).

The cannabinoid content of each plant part varies, paralleling observable gland distributions (Fetterman, Keith et al., 1971; Honma et al., 1971a, 1971b; Kimura and Okamoto, 1970; Ohlsson et al., 1971; Ono, Shimamine, and Takahashi, 1972), although Turner, Hemphill, and Mahlberg (1978) have disagreed. Roots contain only trace amounts. Stalks, branches, and twigs have greater quantities, although not as much as leaf material. Vegetative leaf contains varying quantities depending on its position on the plant: lower leaves possessing less and upper ones more. Leaf glands are most dense on the abaxial (underside) surface. The greatest amount of cannabinoids are found in the new growth near each apical tip (Kimura and Okamoto, 1970;

Figure 2.2. Resin-producing stalked glandular trichome.

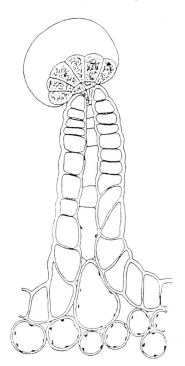

Source: Briosi, G. and Tognini, F. (1894). "Anatomia della canapa. Parte prima: Organi sessuali." *Atti dell' Istituto Botanico della Universita de Pavia Serie II,* 3:91-209.

Steinberg et al., 1975), although Ono, Shimamine, and Takahashi (1972) seem to differ on this point. This variation in leaf gland placement may be due to either loss of glands as the leaf matures or a greater endowment of glands on leaves successively produced as the plant matures. Additional study on this point is required.

Once sexual differentiation has occurred, the generation of female reproductive organs and their associated bracts increases total plant cannabinoid content. Bracts subtending the female flowers contain a greater density of glands than the leaves. The small cuplike bracteole (perigonal bract) enclosing the pistil has the highest cannabinoid content of any single plant part (Kimura and Okamoto, 1970; Honma et al., 1971a, 1971b). Second only to this is the flower itself (Fetterman, Keith et al., 1971). Since it has no reported epidermal gland structures, the cannabinoids present must be due to either undiscovered production sites or simple adherence of resin from the inner surface of its intimately associated bracteole. This conjecture is supported by the finding that the achenes do not contain substantial amounts of the cannabinoids (Fetterman, Keith et al., 1971; Ono, Shimamine, and Takahashi, 1972). Reproductive structures of the male plant are also provided with greater concentrations of the cannabinoids (Fetterman, Keith et al., 1971; Ohlsson et al., 1971). Stalked glands have been observed covering the tepal, with massively stalked glands occurring on the stamen filament (Dayanadan and Kaufman, 1976). In addition, rows of very large sessile glands are found situated in grooves on the anther itself (Dayanadan and Kaufman, 1976; Fairbairn, 1972) and apparently provide the pollen with a considerable cannabinoid content (Paris, Boucher, and Cosson, 1975).

CANNABINOIDS AND ENVIRONMENTAL STRESS

Desiccation

Delta-9-tetrahydrocannabinol (THC) is a viscous hydrophobic oil (Garrett and Hunt, 1974) that resists crystallization (Gaoni and Mechoulam, 1971) and cannabinoids are of low volatility (Adams et al., 1941). Since the sticky resins produced and exuded on the surface of the plant are varying combinations of THC, other cannabinoids, and a variety of terpenes, they can be seen as analogous to the waxy coatings of the cacti and other succulents that serve as a barrier to water loss in dry environments.

Bouquet (1950) has mentioned that the western side of Lebanon's mountainous *Cannabis* growing area is less favorable for resin production because of humid sea winds. De Faubert Maunder (1976) also observed that the copious separable resin needed for hashish production occurs only "in a belt passing from Morocco eastwards, taking in the Mediterranean area, Arabia, the India subcontinent and ending in Indo-China" (pp. 78-79). These are mostly areas notable for their sparse rainfall, low humidity, and sunny climate. Is it merely coincidence that resin is produced according to this pattern, as well?

Experimental evidence is accumulating that reinforces these notions. Sharma (1975) reported a greater glandular trichome density on leaves of *Cannabis* growing in xeric circumstances. Paris, Boucher, and Cosson (1975) have demonstrated a marked increase in the cannabinoid content of *Cannabis* pollen with decreased humidity. Murari et al. (1983) grew a range of *Cannabis* fiber cultivars in three climatic zones of Italy and found higher THC levels in those plants grown in the drier "continental" (versus "maritime") climate. Hakim, El Kheir, and Mohamed (1986) report that CBD-rich English *Cannabis* devoid of THC produced significant amounts of THC and less CBD, when grown in the Sudan. This trend was accentuated in their next generation of plants.

Haney and Kutscheid (1973) have shown significant correlations of plant cannabinoid content with factors affecting soil moisture availability: content of clay or sand, percent slope of plot, and competition from surrounding vegetation. In some cases, this last factor was noted to have induced a stunted plant with "disproportionately smaller roots," tending to increase both the frequency and severity of desiccation stress.

In a study of ten Kansas locations, Latta and Eaton (1975) found wide differences in plant cannabinoid content, observing that "*delta*-9-THC ranged from 0.012 to 0.49 percent and generally increased as locations became less favorable for plant growth, suggesting increased plant stress enhanced *delta*-9-THC production" (p. 161). Mention was also made of a positive correlation between competing vegetation and THC content. Although the sampling area was not considered moisture deficient, they speculated that "Greater difference among locations might have been observed under drought conditions" (p. 161).

Temperature

Temperature may play a role in determining cannabinoid content, but perhaps only through its association with moisture availability. Boucher et al. (1974) reported an increase in cannabinoid content with temperature (32°C versus 22°C), however, some variables such as increased water loss due to accelerated evaporation and plant transpiration at high temperatures were left unaccounted. In contrast, Bazzaz et al. (1975), using four *Cannabis* ecotypes of both tropical and temperate character, demonstrated a definite decrease in cannabinoid production with increased temperature (32°C versus 23°C). Later studies by Braut-Boucher (1980) on clones of two strains from South Africa revealed a more complex pattern of biosynthesis according to strain, gender, and chemical homolog produced. Clearly, further study of this parameter is needed.

Soil Nutrients

Mineral balance seems to influence cannabinoid production. Krejci (1970) found increases related to unspecified "poor soil conditions." Haney and Kutcheid (1973) have shown the influence of soil K, P, Ca, and N concentrations on Illinois *Cannabis*. They report a distinctly negative correlation between soil K and plant *delta*-9-THC content, although K-P interaction, N, and Ca were positively correlated with it. These minerals were also shown to affect the production of CBD, *delta*-8-THC and cannabinol (CBN), although the latter two compounds are now thought to be spontaneous conversion products of *delta*-9-THC. Kaneshima, Mori, and Mizuno (1973) have demonstrated the importance of optimal Fe levels for plant synthesis of THC. Latta and Eaton (1975) reported Mg and Fe to be important for THC production, suggesting that these minerals may serve as enzyme cofactors. Coffman and Gentner (1975) also corroborated the importance of soil type and mineral content, and observed a significant negative correlation between plant height at harvest and THC levels. Interestingly, Marshman, Yawney, and Popham (1976) report greater amounts of THC in Jamaican plants growing in "organically" enriched (versus artificially fertilized) soils.

Insect Predation

Wounding of the plant has been employed as a method to increase resin production (Emboden, 1972). This increase may be a response to desiccation above the point of vascular disruption. Under natural circumstances, wounding most often occurs as a result of insect attack. This is a source of environmental stress that the production of terpenes and cannabinoids may be able to minimize. *Cannabis* is subject to few predators (Smith and Haney, 1973; Stannard, Dewitt, and Vance, 1970) and has even been utilized in powdered or extract form as an insecticide (Bouquet, 1950) or repellent (Khare, Gupta, and Chandra, 1974). Its three apparent defensive mechanisms are a generous covering of nonglandular trichomes, emission of volatile terpenoid substances, and exudation of the sticky cannabinoids. *Cannabis* is often noted for its aromatic quality and many of the terpenes produced are known to possess insect-repellent properties. Among these are *alpha* and *beta* pinene, limonene, terpenoid, and borneol. Pinenes and limonene comprise over 75 percent of the volatiles detected in the surrounding atmosphere, but account for only 7 percent of the essential oil (Hood, Dames, and Barry, 1973). Consistent with glandular trichome density and cannabinoid content, more of these terpenes are produced by the inflorescences than the leaves, and their occurrence is also greater in the female plant (Martin, Smith, and Farmilo, 1961).

No insect toxicity studies using pure cannabinoids have been published to date. Rothschild, Rowen, and Fairbairn (1977) found THC-rich Mexican *Cannabis* fatal to tiger moth *(Arctia caja)* larvae and to one species of grasshopper *(Schistocerca gregaria)*, but not another *(Zonocerus elegans)* from Nigeria. In contrast, they reported that no fatalities resulted from the use of CBD-rich Turkish *Cannabis* as insect food. However, no attempt was made to isolate the particular toxin (presumed THC) responsible for these effects, although insect cannabinoid levels were carefully measured. Rothschild and Fairbairn (1980) later found that aqueous extracts (containing undefined "volatiles") of these *Cannabis* strains, as well as ethanolic tinctures of pure CBD and THC, do act as repellents, affecting oviposition in the large white cabbage butterfly *(Pieris brassicae)*.

The cannabinoids may also serve as a purely mechanical defense. A tiny creature crossing the leaf surface could rupture the tenuously attached globular resin reservoirs of the glandular trichomes (Ledbetter and Krikorian, 1975) and become ensnared in resin. A sizable chewing insect, if able to overcome these defenses, would still have difficulty chewing the gummy resin, along with the cystolithic trichomes and silicified covering trichomes also present on the leaf. The utility of these epidermal features as insect antifeedants is also inferable from their predominant occurrence on the insect-favored abaxial leaf surface. Although the above strategies represent a seemingly sophisticated antifeedant system, many plants (Levin, 1973) and even arthropods (Eisner, 1970) utilize similar defense mechanisms, both often employing terpenes identical to *Cannabis*.

Competition

Terpenes may also help to suppress the growth of surrounding vegetation (Muller and Hauge, 1967; Muller, Muller, and Haines, 1964). Haney and Bazzaz (1970) speculated that such a mechanism may be operative in *Cannabis*. They further ventured that since the production of terpenes is not fully developed in very young plants, this may explain their inability to compete successfully with other vegetation until more mature. The aforementioned observation (Latta and Eaton, 1975) of increased THC production by plants in competition with surrounding vegetation "at a time in the growing season when moisture was not limiting" (p. 161), may indicate a stimulus for cannabinoid production beyond that of simple water stress.

Bacteria and Fungi

The cannabinoids may serve as a protectant against microorganisms. *Cannabis* preparations have long served as medicines (apart from their psychoactive properties) and are effective against a wide variety of infectious diseases (Kabelic, Krejci, and Santavy, 1960; Mikuriya, 1969). These antibiotic properties have been demonstrated with both *Cannabis* extracts (Ferenczy, Grazca, and Jakobey, 1958; Kabelic, Krejci, and Santavy, 1960; Radosevic, Kupinic, and Grlic, 1962) and a variety of isolated cannabinoids (ElSohly et al., 1982; Farkas and Andrassy, 1976; Gal and Vajda, 1970; van Klingeren and ten Ham, 1976). CBG has been compared (Mechoulam and Gaoni, 1965) in both "structure and antibacterial properties to grifolin, an antibiotic from the basidiomycete *Grifolia conflens*" (p. 1228). Ferency (1956) has demonstrated the antibiotic properties of *Cannabis* seed, a factor that may aid its survival when overwintering. Adherent resin on the seed surface, as well as a surrounding mulch of spent *Cannabis* leaves, may serve in this regard.

Some of the many fungal pathogens that affect *Cannabis* include *Alternaria alterata* (Haney and Kutsheid, 1975), *Ascochyta prasadii* (Shukla and Pathak, 1967), *Botryosphaeria marconii* (Charles and Jenkins, 1914), *Cercospora cannabina* and *C. Cannabis* (Lentz et al., 1974), *Fusarium oxysporum* (McCain and Noviello, 1985), *Phoma* sp. (Srivastava and Naithani, 1979) and *Phomopsis ganjae* (McPartland, 1984). Although *A. alterata* attacks Illinois *Cannabis* and destroys 2.8 to 45.5 percent of the seed (Haney and Kutsheid, 1975), the balance of these species are leaf spot diseases. McPartland (1984) has demonstrated the inhibitory effects of THC and CBD on *Phomopsis ganjae*. However, de Meijer, van der Kamp, and van Eeuwijk (1992), in evaluating a large collection of *Cannabis* genotypes, did not find a correlation between cannabinoid content and the occurence of *Botrytis*. Fungal evolution of a mechanism for overcoming the plant's cannabinoid defenses may be responsible for their success as pathogens. Indeed, some have been demonstrated to metabolize THC and other cannabinoids (Binder, 1976; Binder and Popp, 1980; Robertson, Lyle, and Billets, 1975).

Ultraviolet Radiation

Another stress to which plants are subject results from their daily exposure to sunlight. While necessary to sustain photosynthesis, natural light contains biologically destructive ultraviolet radiation. This selective pressure has apparently affected the evolution of certain defenses, among them, a chemical screening functionally analogous to the pigmentation of human skin. A preliminary investigation (Pate, 1983) indicated that, in areas of high ultraviolet radiation exposure, the UV-B (280 to 315 nm) absorption properties of THC may have conferred an evolutionary advantage to *Cannabis* capable of greater production of this compound from biogenetic precursor CBD. The extent to which this production is also influenced by environmental UV-B induced stress has been experimentally determined by Lydon, Teramura, and Coffman (1987). Their experiments demonstrate that under conditions of high UV-B exposure, drug-type *Cannabis* produces significantly greater quantities of THC. They have also demonstrated the chemical lability of CBD upon exposure to UV-B (Lydon and Teramura, 1987), in contrast to the stability of THC and CBC. However, studies by Brenneisen (1984) have shown only a minor difference in

UV-B absorption between THC and CBD, and the absorptive properties of CBC proved considerably greater than either. Perhaps the relationship between the cannabinoids and UV-B is not so direct as first supposed. Two other explanations must now be considered. Even if CBD absorbs on par with THC, in areas of high ambient UV-B the former compound may be more rapidly degraded. This could lower the availability of CBD present or render it the less energetically efficient compound to produce by the plant. Alternatively, the greater UV-B absorbency of CBC compared to THC and the relative stability of CBC compared to CBD might nominate this compound as the protective screening substance. The presence of large amounts of THC would then have to be explained as merely an accumulated storage compound at the end of the enzyme-mediated cannabinoid pathway. However, further work is required to resolve the fact that Lydon's (1985) experiments did not show a commensurate increase in CBC production with increased UV-B exposure.

EVOLUTION OF BIOGENETIC PATHWAYS

This CBC pigmentation hypothesis might imply the development of an alternative to the accepted biochemical pathway from CBG to THC via the photolabile CBD. Until 1973 (Turner and Hadley, 1973), separation of CBD and CBC by gas chromatography was difficult to accomplish, so that many peaks identified as CBD in the preceding literature may in fact have been CBC. Indeed, it has been noted (De Faubert Maunder, 1970) and corroborated by GC/MS (Turner and Hadley, 1973) that some tropical drug strains of *Cannabis* do not contain any CBD at all, yet have an abundance of THC. This phenomenon has not been observed for northern temperate varieties of *Cannabis*. Absence of CBD has led some authors (De Faubert Maunder, 1970; Turner and Hadley, 1973) to speculate that another biogenetic route to THC is involved.

Facts scattered through the literature do indeed indicate a possible alternative. Holley, Hadley, and Turner (1975) have shown that Mississippi-grown plants contain a considerable content of CBC, often in excess of the CBD present. In some examples, either CBD or CBC was absent, but in no case were plants devoid of both.

Their analysis of material grown in Mexico and Costa Rica served to accentuate this trend. Only one example actually grown in their respective countries revealed the presence of any CBD, although appreciable quantities of CBC were found. The reverse seemed true as well. Seed from Mexican material devoid of CBD was planted in Mississippi and produced plants containing CBD. Radioisotope tracer studies (Shoyama et al., 1975) have uncovered the intriguing fact that radiolabeled CBG fed to a very low THC-producing strain of *Cannabis* is found as CBD, but when fed to high THC-producing plants, appeared only as CBC and THC. Labeled CBD fed to a Mexican example of these latter plants likewise appeared as THC, although this may have been an experimental artifact (Shoyama, 1997). Their research also indicated that incorporation of labeled CBG into CBD or CBC was age dependent. Later work (Shoyama, Hirano, and Nishiaka, 1984) indicated that conversion of CBG to CBC was enzymatically controlled, even though the resulting product (having an asymmetric center) is not optically active. Vogelman, Turner, and Mahlberg (1988) likewise reported that the developmental stage of *Cannabis* seedlings, as well as their exposure to light, affects the occurrence of CBG, CBC or THC in Mexican *Cannabis*. Definitive evidence for the enzymatic production of THC directly from CBG has been provided by the results of Taura, Morimoto, Shoyama, and Mechoulam (1995). They isolated an enzyme from Mexican *Cannabis* that was demonstrated to perform this conversion *in vitro*. Also shown was an inability of the enzyme to process either the non-carboxylated compound or CBDA. This group later isolated two other enzymes that converted CBGA to CBDA (Taura, Morimoto, and Shoyama, 1996) and CBGA to CBCA (Morimoto, Komatsu, Taura, and Shoyama, 1997), respectively.

A broad survey of the genus is now needed in order to determine if CBDA can be eliminated as a possible biogenetic precursor of THCA. For the present, these results imply that two distinct biochemical systems for THCA biosynthesis may have evolved: the apparent low-efficiency original CBDA-mediated path of temperate origin plants, and an additional high-efficiency pathway directly from CBGA in tropical strains. These latter chemotypes can also produce CBCA from CBGA at the expense of CBDA, although occasionally the first compound is not found in significant amounts. The entire absence of CBDA in a few of these strains may reflect a loss of the original pathway. The variable CBDA and relatively modest THCA levels found in temperate *Cannabis* may be caused by differing amounts of CBDA-forming and "CBDA-cyclase" enzymes, respectively. It may also be possible that the CBCA-forming enzyme is missing entirely in temperate varieties, since the presence of trace amounts of CBCA can be explained by spontaneous conversion of CBGA (Mechoulam, 1997). This overall pattern of circumstance adds weight to the hypothesis that cannabinoids and their associated terpenes provide a survival advantage to the plant, particularly in the tropical biome.

CONCLUSION

Although *Cannabis* has come under intensive investigation, more work is needed to probe the relationship of its chemistry to biotic and abiotic factors in the environment. Glandular structures on the plant are production sites for the bulk of secondary compounds present and seem placed according to the survival and reproductive priorities of the plant, with new leaf growth and flower-associated bracts being the most populated. It is probable that the cannabinoids and associated terpenes serve as defensive agents in a variety of antidessication, antimicrobial, antifeedant, and UV-B pigmentation roles. UV-B selection pressures may be not only responsible for evolutionary selection of the high-THC/CBC *Cannabis* varieties found in equatorial regions, but also for a new biogenetic pathway directly from CBG to THC in some of these strains. Although environmental stresses appear to be a direct stimulus for enhanced chemical production by individual plants, it must be cautioned that such stresses may also skew data by hastening development of the cannabinoid-rich flowering structures. Careful and representative sampling must be performed to obtain meaningful results. Further investigations are needed to determine possible ecological roles for these compounds and the specific evolutionary advantages they may confer.

REFERENCES

Abel, E. (1980). *Marihuana: The first 12,000 years.* New York: Plenum Press, p. 100.
Adams, R., Caine, C. K., McPhee, W. D., and Wearn, T. N. (1941). "Structure of cannabidiol. XII. Isomerization to tetrahydrocannabinols." *Journal of the American Chemical Society,* 63:2209-2213.
Adams Jr., T. C. and Jones, L. A. (1973). "Long chain hydrocarbons of *Cannabis* and its smoke." *Agricultural and Food Chemistry,* 21:1129-1131.
Bazzaz, F. A., Dusek, D., Seigler, D. S., and Haney, A.W. (1975). "Photosynthesis and cannabinoid content of temperate and tropical populations of *Cannabis sativa.*" *Biochemical Systematics and Ecology,* 3:15-18.
Binder, M. (1976). "Microbial transformation of (-)-*delta*-3,4-*trans*-tetrahydrocannabinol by *Cunninghamella blakesleena* Lender." *Helvetica Chimica Acta,* 63: 1674-1684.
Binder, M. and Popp, A. (1980). "Microbial transformation on cannabinoids. Part 3: Major metabolites of (3R, 4R)-*delta*-l-tetrahydrocannabinol." *Helvetica Chimica Acta,* 2515-2518.
Boucher, F., Cosson, L., Unger, J., and Paris, M. R. (1974). "Le *Cannabis sativa* L.; races chemiques ou varietes." *Plantes Medicinales et Phytotherapie,* 8:20-31.
Bouquet, J. (1950). *Cannabis: UN Bulletin on Narcotics,* 2:14-30.
Braut-Boucher, F. (1980). "Effet des conditions ecophysiologiques sur la croissance, le developpement et le contenu en cannabinoides de clones correspondant aux deux types chimiques du *Cannabis sativa* L. originaire d'Afrique du Sud." *Physiologie Vegetale,* 18:207-221.
Brenneisen, R. (1984). "Psychotrope Drogen II. Bestimmung der Cannabinoid in *Cannabis sativa* L. und in *Cannabis*produkten mittels Hochdruckflussigkeitschromatographie (HPLC)." *Pharmaceutica Acta Helvetiae,* 59:247-259.
Briosi, G. and Tognini, F. (1894). "Anatomia della canapa. Parte prima: Organi sessuali."*Atti dell' Istituto Botanico della Universita de Pavia Serie II,* 3:91-209.
Charles, V. and Jenkins, A. (1914). "A fungous disease of hemp." *Journal of Agricultural Research,* 3:81-85.
Coffman, C. B. and Gentner, W. A. (1975). "Cannabinoid profile and elemental uptake of *Cannabis sativa* L. as influenced by soil characteristics." *Agronomy Journal,* 67:491-497.
Dayanandan, P. and Kaufman, P. B. (1976). "Trichomes of *Cannabis sativa* L. *(Cannabaceae).*" *American Journal of Botany,* 63:578-591.
De Faubert Maunder, M. T. (1970). "A comparative evaluation of the tetrahydrocannabinol content of *Cannabis* plants." *Journal of the Association of Public Analysts,* 8:42-47.
De Faubert Maunder, M. T. (1976). "The forensic significance of the age and origin of *Cannabis.*" *Medicine, Science and the Law,* 16:78-89.
De Pasquale, A., Tumino, G. and De Pasquale, R. C. (1974). "Micromorphology of the epidermic surfaces of female plants of *Cannabis sativa* L." *UN Bulletin on Narcotics,* 26:27-40.

De Zeeuw, R. A., Malingre, T. M., and Merkus, F. W. H. M. (1972). "Tetrahydrocannabinolic acid, an important component in the evaluation of *Cannabis* products." *Journal of Pharmacy and Pharmacology*, 24:1-6.

De Zeeuw, R. A., Vree, T. B., Breimer, D. D., and van Ginnekin, C. A. M. (1973). "Cannabivarichromene, a new cannabinoid with a propyl side chain in *Cannabis*." *Experientia*, 29:260-261.

De Zeeuw, R. A., Wijsbek, J., Breimer, D. D., Vree, T. B. van Ginneken, C. A., and van Rossum, J. M. (1972). "Cannabinoids with a propyl side chain in *Cannabis*. Occurrence and chromatographic behavior." *Science*, 175:778-779.

De Zeeuw, R. A., Wijsbek, J., and Malingre, Th. M. (1973). "Interference of alkanes in the gas chromatographic analysis of *Cannabis* products." *Journal of Pharmacy and Pharmacology*, 25:21-26.

Eisner, T. (1970). "Chemical defense against predation in arthropods." In E. Sondheimer and J. B. Simone (Eds.). *Chemical ecology*, p. 127. New York: Academic Press.

ElSohly, H., Turner, C. E., Clark A. M., and ElSohly, M. A. (1982). "Synthesis and antimicrobial properties of certain cannabichrome and cannabigerol related compounds." *Journal of the Pharmaceutical Sciences*, 71:1319-1323.

ElSohly, M. A. and Turner, C. E. (1976). "A review of nitrogen-containing compounds from *Cannabis sativa* L." *Pharmaceutisch Weekblad*, III:1069-1075.

Emboden, W. A. (1972). "Ritual use of *Cannabis sativa* L.: A historical-ethnographic survey." In P. Furst (Ed.). *Flesh of the gods*, p. 224. New York: Praeger Press.

Fairbairn, J. W. (1972). "The trichomes and glands of *Cannabis sativa* L." *UN Bulletin on Narcotics*, 24:29-33.

Farkas, J. and Andrassy, E. (1976). "The sporostatic effect of cannabidiolic acid." *Acta Alimentaria*, 5:57-67.

Fellermeier, M., and Zenk, M. H. (1998). "Prenylation of olive tolate by a hemp transferase yields cannabigerolic acid; the precursor of tetrahydrocannabinol." *Federation of European Biochemical Societies Letters*, 427(2):283-285.

Ferenczy, L. (1956). "Antibacterial substances in seeds of *Cannabis*." *Nature*, 178:639.

Ferenczy, L., Grazca, L., and Jakobey, I. (1958). "An antibacterial preparation from hemp (*Cannabis sativa* L.)." *Naturwissenschaften*, 45:188.

Fetterman, P. S., Doorenbos, N. J., Keith, E. S., and Quimby, M. W. (1971). "A simple gas liquid chromatography procedure for determination of cannabinoidic acids in *Cannabis sativa* L." *Experientia*, 27:988-990.

Fetterman, P. S., Keith, E. S., Waller, C. W., Guerrero, O., Doorenbos, N. J., and Quimby, M. W. (1971). "Mississippi-grown *Cannabis sativa* L.: Preliminary observation on chemical definition of phenotype and variations in tetrahydrocannabinol content versus age, sex, and plant part." *Journal of the Pharmaceutical Sciences*, 60:1246-1249.

Fetterman, P. S. and Turner, C. E. (1972). "Constituents of *Cannabis sativa* L. I. Propyl homologs of cannabinoids from an Indian variant." *Journal of the Pharmaceutical Sciences*, 61:1476-1477.

Gal, I. E. and Vajda, O. (1970). "Influence of cannabidiolic acid on microorganisms." *Elelmezesi Ipar,* 23:336-339.

Gaoni, Y. and Mechoulam, R. (1971). "The isolation and structure of *delta*-1-tetrahydrocannabinol and other neutral cannabinoids from hashish." *Journal of the American Chemical Society,* 93:217-224.

Garrett, E. R. and Hunt, C. A. (1974). "Physico-chemical properties, solubility and protein binding of *delta*-9-tetrahydrocannabinol." *Journal of the Pharmaceutical Sciences,* 63:1056-1064.

Gellert, M., Novak, I., Szell, M., and Szendrei, K. (1974). Glycosidic components of *Cannabis sativa* L. I. Flavonoids. UN Document ST/SOA/SER.S/50 September 20.

Gill, E. W. (1971). "Propyl homologue of tetrahydrohcannabinol: Its isolation from *Cannabis*, properties and synthesis." *Journal of the Chemical Society,* 579-582.

Gill, E. W., Paton, W. D. M., and Pertwee, R. G. (1970). "Preliminary experiments on the chemistry and pharmacology of *Cannabis*." *Nature,* 228:134-136.

Grlic, L. and Andrec, A. (1961). "The content of acid fraction in *Cannabis* resin of various age and provenance." *Experientia,* 17:325-326.

Hakim, H. A., El Kheir, Y. A., and Mohamed, M. I. (1986). "Effect of climate on the content of a CBD-rich variant of *Cannabis*." *Fitoterapia,* 57:239-241.

Hammond, C. T. and Mahlberg, P. G. (1973). "Morphology of glandular hairs of *Cannabis sativa* from scanning electron microscopy." *American Journal of Botany,* 60:524-528.

Hammond, C. T. and Mahlberg, P. G. (1978). "Ultrastructural development of capitate glandular hairs of *Cannabis sativa* L. (*Cannabaceae*)." *American Journal of Botany,* 65:140-151.

Haney, A. and Bazzaz, F. A. (1970). Discussion in C. R. B. Joyce and S. H. Curry, (Eds.). *The botany and chemistry of Cannabis.* London: Churchill.

Haney, A. and Kutscheid, B. B. (1973). "Quantitative variation in chemical constituents of marihuana from stands of naturalized *Cannabis sativa* L. in east central Illinois." *Economic Botany,* 27:193-203.

Haney, A. and Kutscheid, B. B. (1975). "An ecological study of naturalized hemp (*Cannabis sativa* L.) in east-central Illinois." *American Midland Naturalist,* 93: 1-24.

Hanus, I. (1975a). "The present state of knowledge in the chemistry of substances of *Cannabis sativa* L. IV. Nitrogen containing compounds." *Acta Universitatis Palackianae Olomucensis Facultatis Medicae,* 73:241-244.

Hanus, I. (1975b). "The present state of knowledge in the chemistry of substances of *Cannabis sativa* L. III. Terpenoid substances." *Acta Universitatis Palackianae Olomucensis Facultatis Medicae,* 73:233-239.

Hanus, I. (1976a). "The present state of knowledge in the chemistry of substances of *Cannabis sativa* L. V. Addendum to part I-IV." *Acta Universitatis Palackianae Olomucensis Facultatis Medicae,* 76:153-166.

Hanus, I. (1976b). "The present state of knowledge in the chemistry of substances of *Cannabis sativa* L. VI. The other contained substances." *Acta Universitatis Palackianae Olomucensis Facultatis Medicae,* 76:167-173.

Harvey, O. J. (1976). "Characterization of the butyl homologs of *delta*-1-tetrahydrocannabinol and cannabidiol in samples of *Cannabis* by combined gas chromatography and mass spectrometry." *Journal of Pharmacy and Pharmacology,* 28:280-285.

Hendricks, H., Malingre, T. M., Batterman, S., and Bos, R. (1975). "Mono- and sesquiterpene hydrocarbons of the essential oil of *Cannabis sativa*." *Phytochemistry,* 14:814-815.

Holley, J. H., Hadley, K. W., and Turner, C. E. (1975). "Constituents of *Cannabis sativa* L. XI. Cannabidiol and cannabichromene in samples of known geographical origin." *Journal of the Pharmaceutical Sciences,* 64:892-895.

Honma, S., Kaneshima, H., Mori, M., and Kitsutaka, T. (1971a). "*Cannabis* grown in Hokkaido. 2. Contents of cannabinol, tetrahydrocannabinol and cannabidiol in wild *Cannabis*." *Hokkaidoritsu Eisei Kenkyushoho,* 21:180-185.

Honma, S., Kaneshima, H., Mori, M., and Kitsutaka, T. (197lb). "*Cannabis* grown in Hokkaido. 3. Variation in the amount of narcotic components of *Cannabis* and its growth." *Hokkaidoritsu Eisei Kenkyushoho,* 21:186-190.

Hood, L. V. S., Dames, M. E., and Barry, G. T. (1973). "Headspace volatiles of marijuana." *Nature,* 242:402-403.

Isbell, H. (1973). "Research on *Cannabis* (marijuana)." *UN Bulletin on Narcotics,* 25:37-48.

Kabelik, J., Krejci, Z., and Santavy, F. (1960). "*Cannabis* as a medicament." *UN Bulletin on Narcotics,* 12:5-23.

Kaneshima, H., Mori, M., and Mizuno, N. (1973). "Studies on *Cannabis* in Hokkaido (Part 6). The dependence of *Cannabis* plants on iron nutrition." *Hokkaidoritsu Eisei Kenkyushoho,* 23:3-5.

Khare, B. P., Gupta, S. B., and Chandra, S. (1974). "Biological efficacy of some plant materials against *Sitophilus oryzae* Linneaeous." *Indian Journal of Agricultural Research,* 8:243-248.

Kimura, M. and Okamoto, K. (1970). "Distribution of tetrahydrocannabinolic acid in fresh wild *Cannabis*." *Experientia,* 26:819-820.

Krejci, Z. (1970). "Changes with maturation in amounts of biologically interesting substances of *Cannabis*." In C. R. B. Joyce and S. H. Curry (Eds.). *The botany and chemistry of Cannabis,* p. 49. London: Churchill.

Lanyon, V. S., Turner J. C., and Mahlberg, P. G. (1981). "Quantitative analysis of cannabinoids in the secretory product from captitate-stalked glands of *Cannabis sativa* L. (Cannabaceae)." *Botanical Gazette,* 142:316-319.

Latta, R. P. and Eaton, B. J. (1975). "Seasonal fluctuations in cannabinoid content of Kansas marijuana." *Economic Botany,* 29:153-163.

Ledbetter, M. C. and Krikorian, A. D. (1975). "Trichomes of *Cannabis sativa* as viewed with scanning electron microscope." *Phytomorphology,* 25:166-176.

Lentz, P. L., Turner, C. E., Robertson, L.W. and Gentner, W.A. (1974). "First North American record for *Cercospora cannabina*, with notes on the identification of *C. cannabina* and *C. cannabis*." *Plant Disease Reporter*, 58:165-168.

Levin, D. A. (1973). "The role of trichomes in plant defense." *Quarterly Review of Biology*, 48:3-16.

Lydon, J. (1985). "The effects of Ultraviolet-B radiation on the growth, physiology and cannabinoid production of *Cannabis sativa* L." PhD dissertation, University of Maryland.

Lydon, J. and Teramura, A. H. (1987). "Photochemical decomposition of cannabidiol in its resin base." *Phytochemistry*, 26:1216-1217.

Lydon, J., Teramura, A. H., and Coffman, C. B. (1987). "UV-B radiation effects on photosynthesis, growth and cannabinoid production of two *Cannabis sativa* chemotypes." *Photochemistry and Photobiology*, 46:201-206.

Mahlberg, P. G. and Kim, E. S. (1992). "Secretory vesicle formation in glandular trichomes of *Cannabis sativa (Cannabaceae)*." *American Journal of Botany*, 79: 166-173.

Malingre, T. N., Hendricks, H., Batterman, S., Bos, R., and Visser, J. (1975). "The essential oil of *Cannabis sativa*." *Planta Medica*, 28:56-61.

Marshman, J., Yawney, C. D., and Popham, R. E. (1976). "A note on the cannabinoid content of Jamaican ganja." *UN Bulletin on Narcotics*, 28:63-68.

Martin, L., Smith, D. M., and Farmilo, C. G. (1961). "Essential oil from fresh *Cannabis sativa* and its use in identification." *Nature*, 191:774-776.

Masoud, A. N., and Doorenbos, N. J. (1973). "Mississippi-grown *Cannabis sativa* L. III. Cannabinoid and cannabinoic acid content." *Journal of the Pharmaceutical Sciences*, 62:313-315.

McCain, A. H. and Noviello, C. (1985). Biological control of *Cannabis sativa*. In Proceedings of the sixth international symposium on biological control of weeds, pp. 635-642. E. S. Delfosse, Ed. Agricultural Canada, Ottawa, Canada.

McPartland, J. M. (1984). "Pathogenicity of *Phomopsis ganjae* on *Cannabis sativa* and the fungistatic effect of cannabinoids produced by the host." *Mycopathologia*, 87:149-154.

Mechoulam, R. (1970). "Marijuana chemistry." *Science*, 168:1159-1166.

Mechoulam, R. (1988). Alkaloids in *Cannabis sativa*. In A. Brossi (Ed.). *The alkaloids*, 34:77-93. San Diego, CA: Academic Press.

Mechoulam, R. (1997). Personal communication.

Mechoulam, R. and Gaoni, Y. (1965). "Hashish IV. The isolation of cannabinolic, cannabidiolic and cannabigerolic acids." *Tetrahedron*, 21:1223-1229.

Merkus, F. W. H. M. (1971). "Two new constituents of hashish." *Nature*, 2332:579-580.

Meijer de, E. P. M., van der Kamp, H. J., and van Eeuwijk, F. A. (1992). "Characterization of *Cannabis* accessions with regard to cannabinoid content in relation to other plant characters." *Euphytica*, 62:187-200.

Mikuriya, T. H. (1969). "Marijuana in medicine: Past, present and future." *California Medicine*, 110:34-40.

Mobarak, Z., Bieniek, D., and Korte, F. (1974a). "Studies on non-cannabinoids of hashish. Isolation and identification of some hydrocarbons." *Chemosphere*, 3:5-8.

Mobarak, Z., Bieniek, D., and Korte, F. (1974b). "Studies on non-cannabinoids of hashish II. An approach to correlate the geographical origin of *Cannabis* with hydrocarbon content by chromatographic analysis." *Chemosphere*, 3:265-270.

Morimoto, S., Komatsu, K., Taura, F., and Shoyama, Y. (1997). "Enzymological evidence for cannabichromenic acid biosynthesis." *Journal of Natural Products*, 60:854-857.

Muller, C. H., Muller, W. H., and Haines, B. L. (1964). "Volatile growth inhibitors produced by aromatic shrubs." *Science*, 143:471-473.

Muller, W. H. and Hauge, R. (1967). "Volatile growth inhibitors produced by *Salvia leucophylla:* Effect on seedling anatomy." *Bulletin of the Torrey Botanical Club*, 94:182-190.

Murari, G., Lombardi, S., Puccini, A. M., and De Sanctis, R. (1983). "Influence of environmental conditions on tetrahydrocannabinol (*delta*-9-THC) in different cultivars of *Cannabis sativa* L." *Fitoterapia*, 54:195-201.

Ohlsson, A., Abou-Chaar, C. I., Agurell, S., Nilsson, I. M., Olofsson, K., and Sandberg, F. (1971). "Cannabinoid constituents of male and female *Cannabis sativa*." *UN Bulletin on Narcotics*, 23:29-32.

Ono, M., Shimamine, M., and Takahashi, K. (1972). "Studies on *Cannabis*. III. Distribution of tetrahydrocannabinol in the *Cannabis* plant." *Eisei Shikenjo Hokoku*, 90:1-4.

Paris, M., Boucher, F., and Cosson, L. (1975). "The constituents of *Cannabis sativa* pollen." *Economic Botany*, 29:245-253.

Paris, R. R., Henri, E., and Paris, M. (1975). "O c-flavonoidima *Cannabis sativa* L." *Arhiv Pharmaciju*, 25:319-328.

Paris, R. R. and Paris, M. R. (1973). "Sur les flavonoides du chanvre (*Cannabis sativa* L.)." *Competes Rendus de l'Academie des Sciences Serie D*, 277: 2369-2371.

Pate, D.W. (1983). "Possible role of ultraviolet radiation in evolution of *Cannabis* chemotypes." *Economic Botany*, 37:396-405.

Radosevic, A., Kupinic, M., and Grlic, L. (1962). "Antibiotic activity of various types of *Cannabis* resin." *Nature*, 195:1007-1009.

Robertson, L. W., Lyle, M. A., and Billets, S. (1975). "Biotransformation of cannabinoids by *Syncephalastrum racemosum.*" *Biomedical Mass Spectroscopy*, 2:266-271.

Ross, S. A. and ElSohly, M. A. (1996). "The volatile oil composition of fresh and air-dried buds of *Cannabis sativa*" *Journal of Natural Products*, 59:49-51.

Rothschild, M. and Fairbairn, J. W. (1980). "Ovipositing butterfly (*Pieris brassicae* L.) distinguishes between aqueous extracts of two strains of *Cannabis sativa* L. and THC and CBD." *Nature*, 286:56-59.

Rothschild, M., Rowen, M. G., and Fairbairn, J. W. (1977). "Storage of cannabinoids by *Arctia caja* and *Zonocerus elegans* fed on chemically distinct strains of *Cannabis sativa*." *Nature*, 266:650-651.

Sharma, G. K. (1975). "Altitudinal variation in leaf epidermal patterns of *Cannabis sativa*." *Bulletin of the Torrey Botanical Club*, 102:199-200.

Shoyama, Y. (1997). Personal communication.
Shoyama, Y., Hirano, H., and Nishioka, I. (1984). "Biosynthesis of propyl cannabinoid acid and its biosynthetic relationship with pentyl and methyl cannabinoids." *Phytochemistry*, 23:1909-1912.
Shoyama, Y., Yagi, M., Nishioka, I., and Yamauchi, T. (1975). "Biosynthesis of cannabinoid acids." *Phytochemistry*, 14:2189-2192.
Shukla, D. D. and Pathak, V. N. (1967). "A new species of *Ascochyta* on *Cannabis sativa* L." *Sydowia Annals of Mycology*, 21:277-278.
Small, E. and Beckstead, H. D. (1973). "Common cannabinoid phenotypes in 350 stocks of *Cannabis*." *Lloydia*, 36:144-165.
Smith, G. E. and Haney, A. (1973). "*Grapholitha tristrigana* (Clemens) *(Lepidoptera: Tortricidae)* on naturalized hemp (*Cannabis sativa* L.) in east-central Illinois." *Transactions of the Illinois State Academy of Sciences*, 66:38-41.
Srivastava, S. L. and Naithani, S. C. (1979). "*Cannabis sativa* Linn., a new host for *Phoma* sp." *Current Science of India*, 48:1004-1005.
Stannard, L. J., Dewitt, J. R., and Vance, T. C. (1970). "The marijuana thrips, *Oxythrips cannabensis*, a new record for Illinois and North America." *Transactions of the Illinois Academy of Sciences*, 63:152-156.
Steinberg, S., Offermeier, J., Field, B. I., and Jansen van Ryssen, F. W. (1975). "Investigation of the influence of soil types, environmental conditions, age and morphological plant parts on the chemical composition of *Cannabis sativa* (Dagga) plants." *South African Medical Journal*, 45:279.
Taura, F., Morimoto, S., and Shoyama, Y. (1996). "Purification and characterization of cannabidiolic-acid synthase from *Cannabis sativa* L." *Journal of Biological Chemistry*, 271(29):17411-17416.
Taura, F., Morimoto, S., Shoyama, Y., and Mechoulam, R. (1995). "First direct evidence for the mechanism of *delta*-1-tetrahydrocannabinolic acid biosynthesis." *Journal of the American Chemical Society*, 117:9766-9767.
Turner, C. E., ElSohly, M. A., and Boeren, E. G. (1980). "Constituents of *Cannabis sativa* L. XVII. A review of the natural constituents." *Journal of Natural Products*, 43:169-234.
Turner, C. E. and Hadley, K. (1973). "Constituents of *Cannabis sativa* L. II. Absence of cannabidiol in an African variant." *Journal of the Pharmaceutical Sciences*, 62:251-255.
Turner, C. E., Hadley K., and Fetterman, P. S. (1973). "Constituents of *Cannabis sativa* L. VI: Propyl homologs in samples of known geographic origin." *Journal of the Pharmaceutical Sciences*, 62:1739-1741.
Turner, C. E., Hadley, K. W., Fetterman, P. S., Doorenbos, N. J., Quimby, M. W., and Waller, C. (1973). "Constituents of *Cannabis sativa* L. IV: Stability of cannabinoids in stored plant material." *Journal of the Pharmaceutical Sciences*, 62:1601-1605.
Turner, J. C., Hemphill, J. K., and Mahlberg, P. G. (1978). "Cannabinoid composition and gland distribution in clones of *Cannabis sativa* L. *(Cannabaceae)*." *UN Bulletin on Narcotics*, 30:55-65.

Turner, J. C. and Mahlberg, P. G. (1988). "In vivo incorporation of labeled precursors into cannabinoids in seedlings of *Cannabis sativa* L. *(Cannabaceae)*." In G. Chesher, P. Consroe, and R. Musty (Eds.). *Marihuana*, pp. 263-270. Canberra: Australian Government Publications.

van Klingeren, B. and ten Ham, M. (1976). "Antibacterial activity of *delta*-9-tetrahydrocannabinol and cannabidiol." *Antonie van Leeuwenhoek Journal of Microbiology and Serology*, 42:9-12.

Vogelman, A. F., Turner, J. C., and Mahlberg, P. G. (1988). "Cannabinoid composition in seedlings compared to adult plants of *Cannabis sativa*." *Journal of Natural Products*, 51:1075-1079.

Vree, T. B., Breimer, D. D., van Gienneken, C. A. M., and Rossum, J. M. (1971). "Identification of methyl and homologs of CBD, THC and CBN in hashish by a new method of combined gas chromatography-mass spectrometry." *Acta Pharmaceutica Suecica*, 8:683-684.

Vree, T. B., Breimer, D. D., van Gienneken, C. A. M., and Rossum, J. M. (1972a). "Identification of cannabicyclol with a pentyl or propyl side-chain by means of combined gas chromatography-mass spectrometry." *Journal of Chromatography*, 74:124-127.

Vree, T. B., Breimer, D. D., van Gienneken, C. A. M., and Rossum, J. M. (1972b). "Identification in hashish of tetrahydrocannabinol, cannabidiol and cannabinol analogs with methyl side-chain." *Journal of Pharmacy and Pharmacology*, 24:7-12.

Chapter 3

Detecting and Monitoring Plant THC Content: Innovative and Conventional Methods

Gianpaolo Grassi
Paolo Ranalli

INTRODUCTION

There are about 400 chemicals present in hemp, with 60 belonging to the family of cannabinoids. The psychoactive molecules that confer the characteristic of drug to hemp are 11-nor-Δ^9-tetrahydrocannabinol and 11-nor-Δ^8-tetrahydrocannabinol (Agurel, Dewey, and Willette, 1984); the only difference is a double bond between the carbon in positions 8 and 9 (see Figure 3.1).

Figure 3.1. Δ^8 and Δ^9-tetrahydrocannabinols.

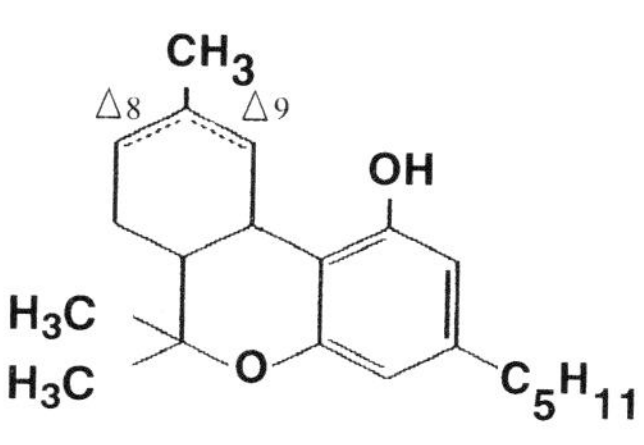

Other cannabinoids such as cannabivarina and tetrahydrocannabivarina have some psychotropic effects, but their hemp content is very limited. Normally, only THC and cannabidiol reach the highest concentrations in hemp as compared with all the others. Analysis of cannabinoids can be performed with traditional methods (GC, HPLC, GC-MS, and HPLC-MS, and TLC) or by immunological methods based on polyclonal or monoclonal antibodies.

Researchers largely borrowed technologies from medicine for the first agrochemical immunoassays and today this trend continues. As regards Δ^9-tetrahydrocannabinol (Δ^9-THC) analysis, the important medical and human involvement has increased the production and spreading of many types of immunoassays, first aimed at detecting the principal metabolite of natural Δ^9-THC which is 11-nor-Δ^9-tetrahydrocannabinol-9-carboxylic acid (Δ^9-carboxy-THC). The simple transfer of immunological tests from the hospital laboratory to agricultural applications is not easy due to problems of matrix interference and the specificity of the antibodies used to perform the analysis; in fact, many polyclonal antibodies and recently monoclonal antibodies, are highly specific for the target molecule, in this case Δ^9-carboxy-THC. Thus the related compounds, Δ^9-THC, Δ^8-THC, cannabidiol (CBD), and cannabinol (CBN) may not be accurately evaluated.

At present, more than ten companies sell polyclonal or monoclonal antibodies against THC (Linscott's Directory, 1995) which all have different specificity (see Table 3.1). The antibodies currently available are not all allowed for diagnostic kit production or there are particular conditions imposed on the use of the antibodies for commercial distribution.

IMMUNODIAGNOSTIC

The immunoassays are based on the mass action law and are all competitive binding assays. What is critical is the binding protein or immunogen, because antibodies must be raised by scheduled immunization using an appropriate antigen. Immunogen preparation is a strategic aspect of antibody production, and for tetrahydrocannabinol in particular many solutions have been attempted (Ullman et al., 1993; Goto, Shima, Morimoto, 1994).

Table 3.1. Cross reactivity of commercial antibodies tested to detect Δ^9-THC used in different immunoassays.

Cannabinoids	Antibody					
	A%	B%	C%	D%	E%	F%
11-nor Δ^9-THC	3.0	100	5	50	10	10
11-nor Δ^8-THC	3.7	0.01	3	—	10	10
11-nor Δ^9-THC-COOH	100	1.0	100	100	100	100
11-nor Δ^8-THC-COOH	236	0.001	98	95	100	84
Cannabidiol (CBD)	0.1	0.01	0.01	0.01	0.001	0.3
Cannabinol (CBN)	0.1	1.0	0.01	0.001	5	4.5
Half displacement or MDC*	$\cong$500	0.0002	5	25	50	0.6

A = Diatech Inc., c-ELISA
B = Pharmadiagnostic Inc., c-ELISA
C = Roche, RIA
D = Abbott, TDx
E = Boehringer, Frontline
F = Biogenesis, c-ELISA
*MDC = Minimal Detectable Concentration (ng/mL of THC)

POLYCLONAL ANTIBODIES

Polyclonal antibody production started a long time ago and this technology continues to develop with a view to replacing and refining the use of laboratory animals. In recent years alternatives have been proposed; for example, the substitution of mammalian antibodies with avian antibodies (ECVAC, 1996). In polyclonal antisera several antibody fractions bind the analyte with different affinities and specificities. The individual antisera differ in their susceptibility to interference on the part of plant substances and the components of the sample tested. To avoid the high cost of standardizing every antisera, the companies use pooled sera that are fully characterized. However, the need for antibodies with constant quality and with the same specificity has increased the interest for advanced techniques to produce monoclonal antibodies and recently recombinant antibodies.

MONOCLONAL ANTIBODIES

In 1975, Koler and Milstein introduced the technique that has changed immunochemistry completely in many fields, and THC has also been the object of attention by the researchers and companies. Many monoclonal antibodies (Mabs) are now available against natural cannabinoid and the metabolites derived following introduction of THC into the human body. The most important advantage of Mabs is that a hybridoma line (immortal cell) that produces an antibody with the necessary specificity and affinity is obtained. The antibody that can be produced is unlimited, thus a test that performs with optimal characteristics could be used forever. The cost of Mab production is quite high, and sometimes, producing Mabs can be difficult (e.g., against THC). Our personal experience is that spleen cells of immunized mice can have a low amount of activated lymphocytes; this is due to proliferation suppression (Nakano, Pross, and Friedman, 1992). The unique epitope recognized by a single Mab sometimes represents a disadvantage, and therefore a pool of Mabs could be used to avoid the risk of an incorrect evaluation due to a small variant in the molecule to be detected. To detect THC in hemp, the higher the Mab specificity for this cannabinoid the better the evaluation of hemp drug content. In fact, the cross-reactivity of Mab anti-THC with other cannabinoids is the biggest problem in testing the THC content in plant samples. For instance, it may be desirable to have a very high affinity antibody for a particular assay, or a low affinity for immunoaffinity purification, or a Mab very resistant to solvent and the matrix effect, for a field test. In conclusion, up until now Mabs have represented the most advanced tool for developing immunoassays with more favorable characteristics than other antibodies.

RECOMBINANT ANTIBODIES

The potential of this new generation of antibodies seems very high. First, cost to produce the antibody could be much lower than monoclonal technology, as well as being lower than the cost of producing a polyclonal antibody. The recombinant antibody is immortal similar to the cells that produce Mabs, and the multiplication technique takes advantage of experience with the bioreactor used for bacteria. The fraction of the antibody molecule that could be obtained with recombinant technologies (single chain fragment variable, scFv, or FAB antibody) is produced by bacterium *Escherichia coli* using very cheap medium. The potential of the recombinant antibody technique is linked to the possibility of manipulating the active polypeptide DNA that corresponds to an IgG fragment with the same affinity and specificity as the entire immunoglobulin. Recombinant DNA technologies could be used to change single amino acids in the antibody fragments to obtain a particular affinity or specificity. Now it is possible to fuse some active parts of proteins, for example, the antibody fragment, with the alkaline phosphatase enzyme (Harper et al., 1997) and some short amino acids such as c-Myc or a histidine tail (Linder et al., 1992) to combine the binding activity of the antibody with an enzyme or marker that can be detected with sensitive systems. Many scientists in the field of immunodiagnostics, however, did not anticipate the complexity of cloning, assembling, and expressing antibody molecules. Despite the difficulty of the task, great progress has been made, although this technology is still far from routine practice.

ASSAY FORMAT

As mentioned before, the immunoassays are all based on competitive binding between a ligand and antigen. To reduce the environmental impact of waste derived from a wide use of isotopes in recent decades the immunoassay has been developed using enzymes and new sensitive substrates.

Radioimmunoassay (RIA)

Originally, radioimmunoassay was the method used to detect small molecules or proteins because its sensitivity was higher than other immunoassay. One kit from Roche (Abuscreen) is available to detect THC metabolites in human urine. In a traditional RIA, which includes a polyclonal antibody as precipitant, a competitive reaction occurs between 11-nor-Δ^9-tetrahydrocannabinol-corboxylic acid-H_3 and the THC metabolite in the sample. The complex of antibody-cannabinoids is precipitated and the result is read with a beta-counter. The radioactivity precipitated is inversely proportional to the THC metabolites present in the sample.

This method is precise and reliable, but the biggest problem is the radioactive tracer, which is included in FDA Class I, and therefore there are many problems with importing. For example, few companies sell it on the Italian market.

Competitive ELISA

When first used, the ELISA format was widely considered to be a qualitative method with limited sensitivity. Starting with environmental chemistry, it was shown that ELISA could have a reproducibility equaling or surpassing RIA with regard to sensitivity. Now this format has been adapted to field applications and so the sum of many essential features has caused the wide diffusion of ELISA in many applications. The method is normally performed in ninety-six well microplates that are coated with antibody (polyclonal or monoclonal) or alternatively, coated with the THC conjugated with a carrier protein. Competitive binding occurs between the THC in the sample and a tracer at a constant concentration, that is, THC conjugated with an enzyme (peroxidase or alkaline phosphatase) in the format with the antibody as coating. When the THC is used as a coating, the antibody reacts with THC in the solution and with the solid phase where THC is blocked. Afterward, the amount of antibody that binds to the coating is detected with an enzyme conjugated anti-IgG. This method is quite simple, fast to perform, and does not require very expensive equipment. There are no problems with reagent waste and sensitivity is good (normally less than 1 ng/mL).

Data can be assessed with the naked eye if it is only necessary to distinguish between a positive or negative sample. Quantitative evaluation of the THC concentrations could be done with the Scachard data plot or a simple computer program. Today, many companies sell complete kits to detect THC in competitive ELISA (Diagnosticx, Inc., Diatech Diagnostic, Hycor Biomedical Inc., Immunotech Corporation, Neogen Corporation). Some of these are specific for tetrahydocannabinol but the majority detect the metabolites and show a low cross reactivity toward natural THC in hemp extracts. We have evaluated some commercial polyclonal and monoclonal antibodies and clone THC-003 from Biogenesis Inc. was used in competitive ELISA to test plant extracts. Its specificity for natural cannabinoids and metabolite is reported in Table 3.1. It was more specific for THC metabolites, and even if these products are not present in hemp extract, it was the cheapest antibody available as an IgG.

The microplates were first coated with a goat anti-IgG antibody and then with monoclonal anti-THC diluted 1:1,000. This procedure was used to increase the sensitivity of the competitive ELISA. A typical standard curve obtained with pure THC as competitor and THC conjugated with peroxidase enzyme as tracer is reported in Figure 3.2.

Figure 3.2. Standard curve obtained with Mab 003 in competitive ELISA using pure Δ^9-THC as competitor.

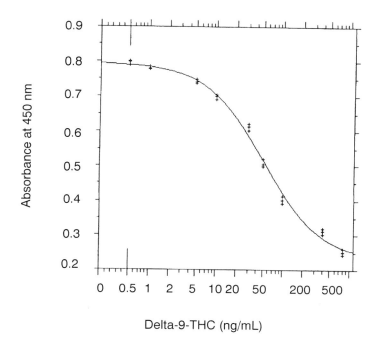

The minimal detectable concentration (MDC) was about 2 ng/mL of pure Δ^9-THC, therefore a dose much lower than the one we have to evaluate in fiber hemp, if we consider that European Community law prescribes a maximum fiber hemp THC content of 0.3 percent in dry weight.

This concentration corresponds to 3,000,000 ng/mL, about 1,000,000 times higher than the MDC of the c-ELISA. We compared the competitive-ELISA in our laboratory, with gas-chromatography analysis using the same hemp extracts (see Figure 3.3).

The correlation coefficient was 0.95; this means that c-ELISA is suitable for the analysis of hemp extracts. The samples tested in c-ELISA were diluted 100 times as compared with the sample injected in the gas-chromatography column. This is because the concentration should be included in the linear zone of the curve (range 13 to 280 ng/mL) for a proper evaluation of hemp THC content.

Figure 3.3. c-ELISA using Mab 003 anti-THC versus gas-chromatography.

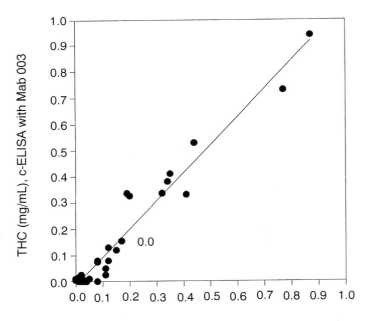

THC (mg/gr d.m.; % d.m.) with gas-chromatography

Competitive ELISA was the method chosen to evaluate THC content of fiber hemp because the number of samples tested in one day was at least ten times that tested with gas-chromatography analysis. The sensitivity of the immunoassay is however the principal aspect. In our breeding program to select new cultivars with very low or zero THC content, the principal aim is to use a very sensitive analytical method. To select the plants with the lowest THC content and to be able to cross the plants, it is necessary to test them when the THC content is differentiated and before blooming, so the analysis must be completed within a short time.

Immuno-Chromatographic Assay

The demand for ready-to-use tests to be performed in a simple laboratory or at home, with particular emphasis on the general human science, has greatly increased the effort to set up very simple, fast, and reliable tests. Starting with the strip to detect pregnancy in women, many devices have been produced. Some very effective tests are now also available to detect THC and its metabolites.

The Contrast kit, SureStep kit, accuPINCH, EZ-Screen, and Frontline are only some of many one-step methods that permit the analysis of THC metabolites in a few minutes. The market for rapid tests to control marijuana abuse is very wide and all the principal diagnostic companies have a patented test. In general, the method is a derivation of a competitive reaction between a monoclonal antibody conjugated or blocked on a chromatographic paper and the antigen (Δ^9-THC-COOH) conjugated with a carrier and fixed in a line or a spot on the same paper. There is also the variant of competitive reaction with an agglutinating support (latex). The result is easily read with the naked eye in few minutes. All these methods are dedicated to the detection of THC metabolites, and very often the natural cannabinoids (cannabinol and cannabidiol) cross-react with the antibody used in these tests. We have tried some of these rapid tests and verified the cross-reactivity claimed by the suppliers, but we were only able to estimate fiber hemp THC content with a new kit, Frontline. This kit was distributed in 1996 by Boehringer Mannheim to test the THC metabolite in human urine, and we adapted the test to hemp extract evaluation. The procedure was very fast, taking only about two minutes to complete. It is based on the principle called GLORIA (Gold Labeled Optically-read

Rapid Immuno Assay). It is only necessary to dip a paper strip into the solution to be tested for three to four seconds, and then wait two minutes to see the color development that is directly proportional to the THC content of the sample. A gold conjugated monoclonal anti-THC is moved to a zone of the paper if the THC is present in the sample. Color references help to verify the Δ^9-THC-COOH concentration of the sample. The cut-off of the test is 50 ng/mL; the positive reaction with Δ^9-THC is obtained with 500 ng/mL. The only problem with the use of Frontline in the field to evaluate hemp plant is the solvent used to extract THC, which is unsuitable for the test and thus it is necessary to dilute the extracts. In our test the readings that corresponded to the data obtained with gas-chromatography analysis were reached when the same concentrated extract used for GC analysis was used diluted 100-fold.

The Frontline test could be used by the police for a preliminary THC content evaluation of fiber hemp crops, to decide directly in the field whether the hemp is really fiber hemp or an illegal drug hemp crop.

Other Immunological Methods

A brief description is given of two other immunological methods very common in hospital laboratories for routine analysis of human urine for THC abuse. These are Enzyme Multiplied Immunoassay (EMIT) (Syva Corporation) and Fluorescence Polarization Immunoassay (TDx) (Abbott Laboratories). The two systems use an automatic machine that performs all the sample dilutions and the incubation steps and gives the final reading after processing the results. The instruments work with an internal standard curve and they seem very specific for THC metabolites.

We have only evaluated directly the TDx method and we found that hemp and also other plant extracts (for example, beet and potato extracts) interfere with the reading of the instrument. The costs of the system and reagents (polyclonal antibodies and a Δ^9-THC-COOH tracer conjugated with a fluorochrome) are quite high. We have no experience with the EMIT method, but we think that the TDx method is unsuitable for wide plant screening and breeding programs that require cheap analysis with simple sample preparation.

Thin Layer Chromatography Assay

Thin Layer Chromatography (TLC) is an analytical method often used in chemistry and was applied in THC metabolite evaluation in human urine (TOXI-MS Cannabinoid Test, Toxi Lab Inc., USA). The rapid method is based on a biphasic thin layer chromatogram combined with a preliminary extraction and concentration of the sample. This method was described by King et al. (1989) and in a comparative work (Foltz and Sunshine, 1990) it was shown that using the TLC method in parallel with the EMIT assay and a reference gas-chromatography mass-spectrometry method with a cut-off of 20 ng/mL, the Δ^9-THC-COOH was correctly identified in 93 percent of samples of the positive urines and 97 percent of the negative urines. In contrast, only 63 percent of the urine specimens shown by GC/MS to contain more than 20 ng/mL of Δ^9-THC-COOH were identified as positive by the EMIT assay at the 100 ng/mL cannabinoid cut-off. We tested TLC analysis using standard plates and Fast Blue BB salt as a detection substrate to evaluate hemp extracts. Using pure cannabinoids it was found that Δ^9-THC was detected by a red spot and other cannabinoids, cannabinol, and cannabidiol were detected by violet or brown spots which showed a quite similar mobility coefficient. The other limitation we found was that the detection threshold is 100 or 200 times higher than competitive ELISA. This method is very simple and does not require particular instruments or expensive equipment if visual reading is used, but for a quantitative evaluation a densitometer is needed and sample loading must be done with very precise equipment. The TLC running in two directions should be a more efficient method to evaluate natural cannabinoids in hemp. However, sensitivity is still the principal limit of this method.

Gas Chromatography (GC) and High Performance Liquid Chromatography (HPLC) Combined with Mass Spectrometry (MS)

All the previous methods described provide only preliminary analytical test results. A more specific alternative chemical method must be used to obtain a confirmed analytical result. Gas chromatography/mass spectrometry has been established at the preferred confirmatory method by the National Institute on Drug Abuse (NIDA). Clinical considerations and professional judgment should be applied to any drug abuse test results, particularly when preliminary positive results are indicated. Hemp evaluation could be done with the same criteria, and when a preliminary and rapid test shows a positive result (THC content higher than 0.3 percent of dry weight) to confirm the analytical result GC/MS or HPLC/MS analysis should be performed. To our knowledge, in Italy, all the laboratories involved in hemp evaluation exclusively for legal drug control, use the traditional methods based on gas-chromatography and sometimes GC in combination with mass-spectrometry. Obviously, this complex protocol greatly limits the number of samples that can be evaluated each day, so in countries where the law restricts the cultivation of fiber hemp the difficulty encountered by the police in controlling all the crops is another reason that prevents or limits the cultivation of this crop.

Gas chromatography, like high performance liquid chromatography methods, requires quite expensive instruments and technicians with particular experience. The sensitivity of the methods is not as high as immunoassay (about 10 or 100 times lower) and solvent waste is a serious problem, at least for HPLC.

Gas chromatography is recommended, however, as the official quantitative method by the European Communities. The original procedure is reported in the Official Bulletin of European Communities, N. 1164/89, annex 5, dated 28-4-1989 (April 28, 1989).

The hemp sample must be taken from a sample of 500 plants randomly harvested in the field when flowering is just finished. The upper third part of the plant should be dried at 40°C in an oven. The seed and stem must be removed and the reduced sample divided into two subsamples, one for the laboratory and the other for counter-analysis, if required.

The sample should be reduced to a fine powder (sieve of 1,000 mesh per cm^2) and 2 g extracted with 40 mL of petroleum ether (40° to 65°C) for twenty-four hours, then shaken for one hour and filtered. The extraction process is carried out twice with the same procedure. The two petroleum portions are pooled and evaporated. The residue is solubilized in 10 mL of petroleum and for quantitative evaluation of THC, 1 mL of extract is dried then solubilized with 2 mL of ethanol with the internal standard of androstene-3-17-dione at 0.1 percent.

A standard curve is set up with five THC concentrations: 0.1, 0.25, 0.50, 1.0, and 1.5 mg/mL with the internal standard at 0.1 percent. The official method suggests the analysis conditions:

oven temperature	240°C
injector temperature	280°C
detector temperature	270°C
carrier flow (helium)	25 mL/min
hydrogen flow	25 mL/min
air flow	300 mL/min

Volume of the sample injected: 1 mL. The concentration is expressed in grams of THC per 100 g of dried sample with a tolerance of 0.03 percent.

The apparatus recommended is a gas chromatograph with a flame ionization detector equipped with a glass column, 2.5 m long and 3.2 mm in diameter, packed with a suitable support impregnated with a stationary phase of phenyl-methyl-silicon.

The modern gas chromatographs mount columns quite different from the previous ones recommended for the analysis of cannabinoids, with a greater sensitivity and separation efficiency. The modern columns used are fused silica, 15 m long and an internal diameter of 0.25 to 0.32 mm. They have an internal thin layer of crossbonded dimethyl-diphenyl polyxiloxane. The volume of the sample injected is 1 µL and the separation temperature can be increased up to 320°C and a suitable internal standard is squalene.

The extraction procedure can be sped up with an ultrasonic instrument or other procedures so that the time required for extraction could be reduced, however traditional THC evaluation should be limited to the confirmatory analysis in those cases in which preliminary tests gave a positive result.

Large Scale THC Analysis

The new interest in fiber hemp in many countries and consequent breeding for new hemp varieties with very low or zero THC content has expanded the demand for large-scale THC analysis. If the law prescribes that every crop must be checked to verify hemp type, the increased police work required for controls indicates that the traditional THC analysis (GC) is not the most suitable method for large-scale testing.

The breeding program to select new varieties must control THC content at every step even if the selection criterion is for fiber or cellulose content. Hundreds or thousands of samples are to be evaluated and, if selection has to be done before blooming starts to isolate the selected plants, but not too early to avoid incorrect evaluation of young plants with low cannabinoid synthesis, the analysis must be performed within a short window of time. The immunological test could satisfy all the demands of a test for hemp breeding and a police control program. They are simple, cheap, sensitive, and require a shorter time than the GC analysis. There are many assay formats and the most appropriate could be chosen for each condition.

Competitive-ELISA, performed in ninety-six well microtiter plates, is the most convenient method for a laboratory that has to deal with hundreds of daily samples. The only limitation in these cases is the extraction procedure. Anti-THC Mabs are available and some companies could supply the tracers (THC conjugated with peroxidase alkaline phosphatase or proteins).

Competitive-ELISA and particular antibodies suffer from the presence of organic solvent in the sample tested. Methanol produces less interference than other solvents, but the concentration must be less than 10 percent of the volume. Some Mabs are stronger than others, however, the lower the solvent level in the sample the better the result. THC extraction from hemp requires solvents, so it is necessary to use methanol or at least dried extract must be solubilized with methanol.

Police controls could be performed directly in the field with a very simple strip test or similar format. The optimization of the extraction procedure for field tests is still in progress. This does not seem to be very problematic because the limit for THC concentration is 0.3 percent in almost all countries and this concentration corresponds to 3,000,000 ng/mL of THC. The cut-off of some strip tests for THC is 50 to 100 ng/mL, therefore the extract sample should be evaluated quite easily.

In our opinion, any strategy to limit spreading of drug hemp field crops must include a very simple and rapid test for a preliminary evaluation of THC concentration. This could provide the police with an efficient method to control the THC content of hemp at all times and everywhere. The number of plants to be tested may not be the same as that indicated in the official method (500 plants), but a randomized sampling of twenty to thirty plants could be an appropriate sample size to estimate the THC content of a hemp crop (de Meijer, van der Kamp, van Eeuwijk, 1992). We tested a hemp sample from 500 plants using GC analysis, and in parallel we tested thirty individual plants out of the 500. The mean THC content evaluated in samples derived from the 500 plants was 0.36 percent of dry weight. The mean value of THC content of a subsample derived from ten plants was 0.35 percent, while the mean of twenty plants was 0.25 percent, and from thirty plants it was 0.26 percent of dry weight. These results also show that the mean THC content of a sample obtained from twenty plants for a preliminary evaluation of hemp crop gave a good estimation of THC concentration.

Variability of THC Content in Hemp

In Italy, studies regarding fiber hemp were stopped in the 1980s because synthetic fibers replaced the traditional vegetable fibers and because the law restricted hemp cultivation. In these years some trials were carried out that attempted to show the effects of growth conditions on THC content in different hemp germplasms. Crops were evaluated in the north and south of Italy and in Sardinia using the European cultivars cultivated at that time: Fibranova, Superfibra, Eletta Campana, Carmagnola, Carmagnola selection, Fibrimon 56, and Fedrina. The THC level, evaluated with GC analysis, was lower than the EC limit of 0.3 percent in all the cultivars. Eletta Campana showed the highest concentration in all the three regions (mean THC content 0.198 percent of dry weight) while Carmagnola and Fibranova showed the lowest THC content (mean THC content 0.044 percent and 0.050 percent respectively). The latitude caused a limited variation in THC content; Emilia Romagna was favorable to a slightly higher THC content than the other two regions.

In 1994, the hemp study started again and the two hemp cultivars (Fibranova and Carmagnola), were evaluated for THC content using GC and immunological methods. The results confirmed that the two Italian cultivars have a THC content below the legal limit, both with values lower than 0.1 percent.

CONCLUSION

The European community, the USA, and other advanced countries currently need to promote alternative crops to avoid surpluses (like cereals). Cultivations with a limited environmental impact are increasingly desirable. Hemp is able to satisfy these requirements, and can provide products (fiber and cellulose) that are in short supply in the EU and other countries, although there is no agreement between different countries about the admissibility of *Cannabis* cultivation, or, where cultivation is allowed, about tolerable cannabinoid levels. Governmental and agricultural organizations can help to introduce regulations that will make it easier for farmers to grow this crop. For *Cannabis* plants to be classified as fiber hemp under the European Union standards, they must not contain more than 0.3 percent THC; the cultivation of the crop must deal with this constraint.

The official method for the detection of THC content in plant tissue is based on gas-chromatographic analysis performed on samples of 500 plants. The spreading of the crop would increase the amount of samples submitted for THC detection. On the other hand, the development of breeding work aimed at developing new hemp varieties requires a series of tests for screening the segregant populations. Thus new methods are urgently needed for the detection of THC.

The official method must be simplified by reducing the size of samples (below 500 plants without losing the reliability of the results) and promoting the use of immunological methods (more sensitive and less expensive). The format and advantages of these methods have been described, as well as the potential they offer for the future.

Currently, these methods are utilized in hospitals to detect *cannabis* metabolites (Δ^9-THC-COOH) in body fluids (blood, urine, etc.). There is no commercially available immunological test to assess natural hemp THC. Thus, a research program was set up three years ago by our group to evaluate commercially available products and to produce monoclonal antibodies against THC. The results obtained have placed us in a pioneering position, since we have raised antibodies successfully and used them for serial tests in breeding programs (Grassi, Faeti, and Ranalli, 1996; Grassi et al., 1997; Grassi, Moschella, and Fiorilli, 1997).

REFERENCES

Agurel, S., Dewey, W. L., and Willette, R. E. (1984). *The cannabinoids: Chemical pharmacologic and therapeutic aspects*. London, New York, San Fransisco: Academic Press.

European Centre for the Validation of Alternative Methods (ECVAC) (1996). *The production of avian (egg yolk) antibodies: IgY*. Alternative to laboratory animals, 24:925-934.

Foltz, R. L. and Sunshine, I. (1990). "Comparison of a TLC method with EMIT and GC/MS for detection of cannabinoids in urine." *Journal of Analytical Toxicology*, 14:375-378.

Goto, Y., Shima, Y., and Morimoto, S. (1994). "Determination of tetrahydrocannabinolic acid-carrier protein conjugate by matrix-assisted laser desorption/ionization mass spectrometry." *Organic Mass Spectrometry*, 29:668-671.

Grassi, G., Faeti, V., and Ranalli, P. (1996). Development of a field test to detect Δ^9-THC (delta-9-tetrahydrocannabinol) in *Cannabis sativa* L. and its employment in selecting genotypes with low content of drug, p. 32. Third European Symposium on "Industrial crops and products," 22-24 April, Reims, France.

Grassi, G., Faeti, V., Moschella, A., and Ranalli, P. (1997). "Indagine sul contenuto di sostanze psicoattive in canapa mediante metodi tradizionali ed immunologici per il miglioramento genetico ed il controllo antidroga." *Sementi Elette*, 2:63-66.

Grassi, G., Moschella, A., and Fiorilli, M. (1997). Evaluation and development of serological methods to detect THC in hemp. Second International Symposium Bioresource Hemp, February 27-March 2, Frankfurt, Germany, pp. 197-201.

Harper, K., Kerschbaumer, R. J. Ziegler, A., Macintosh, S. M., Cowan, G. H., Himmler, G., Mayo, M. A., and Torrance, L. (1997). "A scFv-alkaline phosphatase fusion protein which detects potato leafroll luteovirus in plant extract by ELISA." *Journal of Virological Methods*, 63:237-242.

King, D. L., Gabor, M. J., Martel P. A., and O'Donnell, C. M. (1989). "A rapid sample-preparation technique for thin layer chromatographic analysis for 11-Nor-Δ9-tetrahydrocannabinol-9-carboxylic acid in human urine." *Clinical Chemistry*, 35(1):163-166.

Linder, P., Guth, B., Wuelfing, C., Krebber, C., Steipe, B., Mueller, F., and Plunckthun, A. (1992). "Purification of native proteins from cytoplasm and periplasm of *Escherichia coli* using IMAC and histidine tails: A comparison of proteins and protocols." In *Metal affinity protein separations*, F. H. Arnold (Ed.), Vol. 4, pp. 41-56. London, New York, San Francisco: Academic Press.

Linscott's directory of immunological and biological reagents (1995). Santa Rosa, CA: pp. 1-280.

Meijer de, E. P. M., van der Kamp, H. J., and van Eeuwijk, F. A. (1992). "Characterisation of *Cannabis* accessions with regard to cannabinoid content in relation to other plant characters." *Euphytica*, 62:187-200.

Nakano, Y., Pross, S., and Friedman, H. (1992). Contrasting effect of *delta-9-tetra-hydrocannabinol* on IL-2 activity in spleen and lymph node cells of mice of different age. *Life Sciences,* 52(52):41-51.

Ullman, E. F., Milburn, G., Jelesko, J., Radika, K., Pirio, M., Kempe, T., and Skold, C. (1993). "Anti-immune complex antibodies enhance affinity and specificity of primary antibodies." *Proceedings of the National Academy of Sciences, USA,* 90:1184-1189.

Chapter 4

Agronomical and Physiological Advances in Hemp Crops

Paolo Ranalli

INTRODUCTION

The role of better agronomic practices in exploiting the yield potential of available hemp varieties has been widely recognized. Many aspects of agronomic management of hemp have not progressed since World War II because two problems became outstanding: (1) the large-scale cultivation of cheaper cotton and development of the synthetic fibers industry replaced hemp fiber; (2) the diffusion of the use of hallucinogenic substances (drugs) derived from *Cannabis* and the consequent strict legislation established in all Western countries theoretically forbids the cultivation of Indian hemp, but de facto of any hemp. In countries such as France, Hungary, and the Ukraine, hemp has been cultivated up until the present day and research has also continued. In many other countries, hemp was grown in the old days as a fiber and oilseed crop. Therefore, in spite of the recognition of the utility in knowing the agronomic requirements of field crops by agronomists, the management technology of hemp has not been fully developed and the information on this and related aspects is incomplete.

In general, the former methods cannot be used in modern agriculture because of the high labor and cost demand. Therefore, new technologies have to be developed.

In this chapter a review of the scanty literature on the cultural and other agronomic aspects of hemp is presented. Specific cultural practices vary greatly from place to place and many good combinations of practices are available for wise use of local resources to increase production.

CANNABIS *GENE POOL*

The varietal background needs to be totally updated when the new material exhibiting morphological markers linked to low THC-content is developed, making it possible to distinguish drug-free cultivars by visual inspection. This context includes the efforts to induce new morphological markers concerning the leaves or the stem, easily detectable, linked to low or free content of THC (Faeti, Mandolino, and Ranalli, 1996; Grassi, Moschella, and Fiorilli, 1997; Ranalli et al., 1997). A breeding program based on a mutagenesis approach aimed at this purpose is currently in progress in Italy (Ranalli et al., 1996; Mandolino et al., 1997; Zottini, Mandolino, and Ranalli, 1997).

Up to now, hemp cultivation has been based on a range of cultivars developed in different areas and released for cultivation from time to time. Heterogeneous populations were originally grown throughout the world since they thrive better but are less productive in stressful environments than homogeneous ones. The original populations were selected to obtain remarkably uniform landraces. At the beginning of the twentieth century, Chinese landraces were used to select the now extinct Kentucky cultivar that was cultivated until the mid-1950s in the United States. Carmagnola cultivar was also one of the oldest hemp cultivars grown in European countries, being used for direct production as well as a breeding parent in crossing programs. Thus, it is present in the pedigree of many current cultivars.

In Italy, Carmagnola was improved by mass selection that led to the development of Bolognese, Toscana, and Ferrarese ecotypes (the names provide specifications with respect to the site of cultivation). The predominant breeding approach was continuous mass selection, a cyclic procedure that attempted to upgrade whole populations by directed selection.

As the reservoir of landraces and heterogeneous varieties declined, hybridization was performed to create new variabilities. Half-sib family selection was employed, based on the evaluation of the progeny from each mother plant. This latter is open pollinated and fertilized by more than one father plant. The selection is based on general combining ability, the entire population being used as a tester. The application of the Bredemann method made it possible to choose the male plants before pollination and selection procedures progressed and adapted to different situations.

From the breeding histories it is evident that a considerable mutual genetic relatedness exists among the modern European and West Asian cultivars. Landraces belonging to the Mediterranean and Central Russian fiber hemp ecotype groups and cross-progenies of these two groups have directly been the basis of, or have been used as breeding parent for, each of the present European and West Asian cultivars.

The old cultivars currently available are presented by country.

- *French cultivars* (Fibrimon 21, Fibrimon 24, Fibrimon 56, Fedora 19, Felina 34, Fedrina 74, Futura 77);
- *Hungarian cultivars* (Kompolti, Kompolti Sárgaszárú, Uniko-B, Kompolti Hybrid TC, Fibriko);
- *Polish cultivars* (Bialobrzeskie and Beniko);
- *Italian cultivars* (Carmagnola, CS (Carmagnola Selezionata), Fibranova, Eletta campana, Superfibra); and
- *Cultivars from the former USSR* (Kuban, Zenica, Dneprovskaya Odnodomnaya 6, USO-11, USO-13, USO-15, JUSO-14, YUSO-16, YUSO-31).

Numerous references on agronomic performance, under various cultural treatments in various locations, are available for most of the above-mentioned cultivars. However, as the expression of quantitative agronomic traits strongly depends on the environment, such data cannot simply be pooled in one table.

VARIETY RECOMMENDATIONS

Factors such as yielding ability, the conditions under which the variety has to be cultivated, and the kind of utilization must be considered when recommending a variety. Field trials are currently underway at separate sites in order to assess the comparative performance of cultivars when grown under different environmental conditions. Twenty-four of the described fiber cultivars have been tested simultaneously in standardized trials carried out for the evaluation of the CPRO *Cannabis* germplasm collection in Wageningen, the Netherlands (de Meijer, 1995). Some of the traits involved were: the pattern of phenological development (being related to potential stem and seed production); stem quality (characterized by the fractions of woody core, secondary bark fiber, and primary bark fiber as well as by the length of woody core fibers); contents of the cannabinoids THC and CBD; and resistance to soil pathogens (root-knot nematodes).

Due to the extreme plasticity of some of the tested traits, especially phenological patterns and cannabinoid contents, absolute values apply for the area where the trial is located. However, assuming little interaction between cultivars and latitudes, one can expect that ranking orders of cultivars for most traits are fairly stable.

HEMP IN CROP ROTATION

Within the European Union, the prices of many arable crops have fallen in recent years, and subsidies for food crops have been reduced in an attempt to combat production surpluses. As a result, crop rotation on arable farms in all European countries is increasingly restricted to a few crops that are relatively profitable (potato, sugar beet) and agronomically indispensable in the rotation (cereals). This short crop rotation has increased the incidence of diseases, in particular soil pathogens, and has lowered yields. It has also led to greater use of biocides, in particular, soil fumigants. This is a disturbing development, as it is generally agreed that arable farming should become more sustainable, and use less biocides.

The identification and development of a "new" crop, to be introduced into current rotations, might help solve this problem. Hemp meets these requirements since it improves soil structure, suppresses weeds, and is virtually free from diseases or pests (van der Werf, van Geel, and Wijlhuizen, 1995).

Kok, Coenen, and de Heij (1994) investigated the effect of fiber hemp on three major soil pathogens: the fungus *Verticillium dahliae* and the root-knot nematodes *Meloidogyne chitwoodi* and *Meloido-*

gyne hapla. All three pathogens were suppressed by hemp, and the authors concluded that the introduction of hemp in a crop rotation might improve soil health.

Except in the cases in which plant densities were low (ten or thirty plants m^{-2}), experiments demonstrated that hemp crops effectively suppressed weeds and that no herbicides were needed. This confirms literature reporting such effects, stating that fiber hemp is an extremely efficient weed suppressor and that no herbicides are needed (Tarasov, 1975; Lotz et al., 1991).

SEEDBED

The ideal requirements of seedbed are largely governed by the previously grown crop, the type of soil, the type and intensity of weeds, and whether the crop is grown under irrigated or rain-fed conditions. The crop, however, requires a well-prepared seedbed, free of weeds and other debris. This is necessary to provide a good physical environment for the growing plants and also ensures sufficient capillarity movement of water to the surface.

METHODS OF PLANTING

The hemp is sown either by drilling in the prepared seedbed or by broadcast in the standing paddy. The most prevalent practice is to sow the seed in rows 8 to 15 cm apart with a grain drill or pore at a depth of 2 to 3 cm, covered by a light planking. Drill-sowing is preferred since it ensures even distribution of seed, and uniformity in depth of planting results in better stand and consequently gives higher yields. Deeper sowing affects the yield adversely. The depth of sowing is, however, influenced by the soil type and moisture level. A rainfall after the sowing crusts the soil, preventing plant emergence. Since the seeds of hemp are relatively small, thin stands can result if they are sown too deep. Under favorable moisture conditions, quick emergence is obtained when the seed is sown about 2.5 cm deep. In drier areas where surface moisture is less favorable, it is sometimes necessary to sow at a greater depth to obtain prompt and uniform germination.

THE EFFECT OF TEMPERATURE ON LEAF APPEARANCE AND CANOPY ESTABLISHMENT IN FIBER HEMP

The development and growth of plants and crops can be related to temperature using thermal time, which is the accumulated temperature above a base temperature, the latter being the temperature below which the process being studied apparently stops. In field crops, the course of proportional light interception by the canopy from plant emergence to all ground cover can be described as a logistic function of thermal time (Spitters, 1990). Preliminary results (Meijer et al., 1995) suggested that this approach might be effective in describing canopy establishment in hemp.

Very little quantitative information is currently available on the effect of temperature on the growth and development of hemp (*Cannabis sativa* L.). Tamm (1933) stated that hemp seed needs 96° Cd (base 0° C) for germination and emergence. No base temperatures have been reported for growth and development of hemp. Experiments in controlled conditions and in the field were thus performed to quantify the effect of temperature on the development and growth of fiber hemp (van der Werf, van Geel, and Wijlhuizen, 1995).

Plants were grown in growth chambers at eleven regimes with average temperatures between 10° C and 28° C, and three cultivars were sown in the field in March, April, and May in 1990, 1991, and 1992 in the Netherlands. In the field, thermal time (base 0° C) between sowing and emergence ranged from 68° Cd to 109.5° Cd (average 88.3° Cd): rates of leaf appearance and stem elongation increased linearly with temperature between 10° C and 28° C. The base temperature for leaf appearance was 5.7° C from the growth chamber experiments and 1° C from the field experiments. In the field, the base temperature for the relationship between light interception by the canopy and thermal time was 2.5° C, and thermal time, calculated at the appropriate base temperature, accounted for about 98 percent of the variance in the number of leaves and for 98.6 percent of the variance in the proportion of light intercepted by the canopy. Days from emergence accounted for less of the variance in both parameters than thermal time. Interception of 90 percent of light was attained on average at

465° Cd (base 0° C) after emergence. It can thus be concluded that thermal time is a simple and accurate tool to describe the course of leaf appearance and light interception capacity in fiber hemp (van der Werf, van Geel, and Wijlhuizen, 1995).

SEEDING RATE

Plant density affects the time course of light interception by the canopy and the efficiency of the use of light for dry matter production (Monteith, 1977). Hemp crops are grown at a wide range of plant densities, depending on the goal of production and expected yield level. For seed production a density of thirty plants m^{-2} may be used, whereas for stem or fiber production densities between 50 and 750 plants m^{-2} have been recommended.

The economically optimal plant densities of hemp grown for bast fiber are higher than the lowest plant density that gives maximum stem dry matter yield because plant density continues to improve stem quality when it no longer increases stem yield. With increasing plant density, bast fiber content in the stem tends to increase and the fineness of the bast fibers (an important aspect of fiber quality for spinning purposes) improves (Jakobey, 1965).

With a high plant density the stems do not branch, and a higher fraction of the more valuable bast fiber (primary bast fibers) in the dry matter of the stem is obtained. When the plant has more space, it will develop a wider stalk, with lower bast fibers/wood fibers ratio and with mostly secondary fibers. The secondary fibers are of less quality for textile or papermaking.

Finally, at high plant densities, interplant competition may generate a size hierarchy, i.e., increase variability in the size of the individuals, with large plants suppressing smaller ones. Eventually, self-thinning may occur, if suppressed plants die. Self-thinning is undesirable, as plants that die in the course of the growing season will not be harvested.

In 1991 and 1992, the cultivar Kompolti Hybrid TC was grown at 10, 30, 90, and 270 plants m^{-2} in the Netherlands, at Randwijk (van der Werf et al., 1995). The maximum stem yield (15.1 ton ha^{-1}) and the maximum bark content in the stem (35.7 percent) were obtained at 90 plants m^{-2}. At 270 plants m^{-2}, severe self-thinning caused a lower stem yield (12.9 ton ha^{-1}). Therefore, a density of about 90 plants m^{-2} seems optimal for late-flowering high-yielding fiber hemp crops.

The effect of plant density on light interception by the canopy and on factors affecting light use efficiency, such as appearance and shedding of leaves and flowering date, was also examined (van der Werf, 1997). The hemp cultivar Kompolti Hybrid TC was grown at initial densities of 10, 30, 90, and 270 plants m^{-2} in field experiments on a heavy clay soil in 1991 and 1992 in the center of the Netherlands. Flowering date was defined as the day at which 75 percent of plants carried at least one female flower with visible stigmas or one pedicellate male flower.

In both years, rate of canopy establishment increased with increasing plant density. In 1991 interception of 90 percent of light was attained at 265° Cd (base 2° C) from emergence for the crop grown at 270 plants m^{-2}, at 313° Cd for the crop at 90 plants m^{-2}, at 417° Cd for 30 plants m^{-2} and 537° Cd for 10 plants m^{-2}. In sowing date experiments carried out in 1991, 1992, and 1993 on a peaty sand soil in the northeast Netherlands, 90 percent light interception was attained at 365° Cd (base 2° C) (from emergence). As the crops in these experiments had an initial plant density of 64 plants m^{-2}, their temperature requirement for canopy establishment agrees with the results obtained in the 1991 and 1992 plant density experiments. Maximum light interception was about 99 percent in both years and did not depend on plant density.

Rate of leaf appearance decreased with increasing plant density, for a given density the rate decreased during the growing season. Rate of leaf shedding was much less affected by plant density than rate of leaf appearance. Averaged over the two years, for the densities of 10, 30, 90, and 270 plants m^{-2} respectively 25.9, 23.2, 18.8, and 15.3 leaves were present at 800° Cd (base 1° C) from emergence.

In both years flowering date was later with increasing plant density. For the initial densities of 10, 30, 90, and 270 plants m^{-2} average flowering dates were August 12, 17, 20, and 24, respectively. Regardless of plant density in both years, canopy light interception remained at its maximum of approximately 99 percent until 300° Cd (base 2° C) after flowering, when it started declining, reaching 75 percent at about 650° Cd postflowering.

These experiments have shown major effects of plant density on some key crop characteristics (rate of canopy establishment, rate of leaf appearance, flowering date). Provided the flowering date is known or can be predicted, these data can be used to incorporate the effect of a wide range of plant densities in a simulation model for growth and yield of hemp (van der Werf, 1997).

GROWING CONDITIONS

Hemp is grown worldwide under varied soil and climatic conditions. It adapts to the same climate as wheat. Temperate and cool climatic conditions are suited for its growth. The crop makes its best growth on well-drained, fertile, medium-heavy soils, especially silty loam, clay loam, and silty clays. This crop, like other fibrous plants, is able to extract heavy metals from the soil in amounts higher than many other agricultural crops. Thus hemp can be grown on polluted soil for gradual purification of the soil provided of course, that the emission of metals is stopped (Baraniecki, Grabowski, and Mankowski, 1995).

When hemp is being produced for fiber, it requires a mild, temperate climate, a humid atmosphere, and a rainfall of at least 0.65 m per year, with abundant rain while the seed germinates and until the young plants are well established. Where the plants have been sown close together, they suffer very little damage from wind, rain, and hail. Hemp has low requirements for crop-protecting agents or fertilizers and it fits well in crop rotation.

CULTURAL PRACTICES

Crop management is aimed at optimizing growing conditions. In fact, when the seedlings are about 10 cm high the hoeing and weeding begin. The hemp plant is a good competitor with weeds: with its initial quick growth and expansive leaves, it shades the ground sufficiently and hence weeds have no opportunity to develop.

Hemp is not a particularly hungry crop, requiring the fertility similar to that necessary for wheat, and although some hemp crops will benefit from added nitrogen, this is not always necessary if the soil is truly in good condition.

Hemp grows rapidly and adequate supplies of available nitrogen are essential for satisfactory crops. However, findings available clearly indicate that excessive application should be avoided. In fact, abundant nitrogen causes a more leafy and succulent type of growth and tends to increase stem diameter above the optimum range. The problem is one of adjusting rates and balances of nutrients in such a way as to ensure maximum production and at the same time maintain fiber quality. From the point of view of the grower, some loss in fiber strength may be justified if the increased yields give a greater net income.

Generally, fertilization is based on soil sampling and since literature data on hemp are scarce, phosphate and potassium fertilizers are applied at the recommended rates for highly productive winter wheat crops. Based on experimental results in Italy (Marras and Spanu, 1979), Denmark (Nordestgaard, 1976) and the Netherlands (Aukema and Friederich, 1957), it is estimated that 175 kg N ha^{-1} minus the reserve of soil mineral nitrogen would allow optimal crop growth and yields up to 15 t ha^{-1} of aboveground dry matter.

PHENOLOGICAL DEVELOPMENT

Seeds usually germinate within three to seven days and the seedlings have two seed-leaves (cotyledons) upheld by a relatively long stalk (hypocotyl). The true leaves of the first pair are distinctly stalked, with a narrowly elliptic blade. The leaves of the second pair are much larger, including a long stalk (petiole) with three leaflets radiating from its tip. Subsequent leaves higher on the stem may have as many as eleven leaflets and are arranged alternately (spirally).

The stem is angular and covered with minute hairs. The flowers are produced in great abundance on the upper part of the plant. A flower of *Cannabis* is unisexual, i.e., it is either male (staminate) or female (pistillate). Usually an individual plant bears only one kind of flower, i.e., it is wholly male or wholly female, and the species is accordingly described as dioecious. Male and female flowers may occur on the same plant and this latter is then described as monoecious (Allavena, 1964).

The flower is covered with hairs and glands secreting drops of resin which are produced most abundantly under hot conditions. The ovary has one ovule and the fruit is an achene which contains a single seed.

Phenological events such as seedling emergence, anthesis, and seed maturity demarcate developmental stages of the life cycle and some of them are very important, being associated with stem production.

Flowering date has a major effect on yield potential and is genetically controlled (Hoffmann, 1961). The critical photoperiod (short day) for induction of flowering increases with latitude of adaptation. Many authors describe changes in phenological development when adapted populations are moved to other latitudes. Cultivation at a lower latitude results in earlier anthesis, and cultivation at higher latitude in delayed anthesis (Fleischmann, 1938; Huhnke et al., 1951; Bredemann, Schwanitz, and von Sengbusch, 1957; Hoffmann, 1961).

Day number of anthesis is significantly correlated with length of the vegetative stage. In fact, after flowering, the growth rate of the crop decreases and increasing proportions of the assimilates are used by the inflorescence (Meijer et al., 1995). Stem elongation occurs principally before flowering and the proportion of the stem formed in the generative stage was larger the earlier the accessions started to flower. Very late-flowering accessions reached the ultimate length in the vegetative stage.

In 1990 and 1991, ambient daylength was compared with a twenty-four-hour daylength in field experiments for Fedrina 74 and Kompolti Hybrid TC at Randwijk. The twenty-four-hour daylength did not totally prevent flowering, but did greatly reduce the allocation of dry matter to floral parts. Furthermore, it enhanced the efficiency of post-flowering light use, and increased stem dry matter yield of 2.7 ton ha^{-1} in both cultivars (van der Werf, Haasken, and Wijlhuizen, 1994).

From these experiments, it is clear that stem yield of hemp can be limited by early flowering date. Ideally, a hemp crop grown for stem production should not flower before harvest. Therefore, flowering date is an initial criterion in cultivar choice. In the germplasm evaluations a wide variation was found for day of anthesis and day of seed maturity.

It was concluded that in an efficient crop growth system, seed production and stem production should occur in separate geographic areas, i.e., seed production at a lower latitude and stem production at a higher latitude (de Meijer and Keizer, 1994).

EFFECT OF NITROGEN FERTILIZATION AND ROW WIDTH

In fiber hemp (*Cannabis sativa* L.) a high plant density is desirable, but interplant competition may cause self-thinning, which reduces stem yield and quality. It was investigated whether agronomic factors could reduce self-thinning in hemp. The effects of soil nitrogen level (80 and 200 kg ha^{-1}) and row width (12.5, 25, and 50 cm) were determined in field experiments on self-thinning, growth, yield, and quality of hemp. Soil nitrogen level affected plant morphology before self-thinning occurred. Due to enhanced competition for light, more plants died from self-thinning at 200 than at 80 kg N ha^{-1}. Although dry matter losses resulting from self-thinning were greater at 200 than at 80 kg N ha^{-1}, crop growth rate was greater at 200 than at 80 kg N ha^{-1}. At final harvest in September stem yield of living plants was 10.4 t ha^{-1} at 80 and 11.3 t ha^{-1} at 200 kg N ha^{-1}, bark content in the stem was 35.6 percent at 80 and 34.0 percent at 200 kg N ha^{-1}. The effect of row width on self-thinning was small relative to that of nitrogen level. More self-thinning took place at 50 cm row width than at 12.5 and 25 cm. During early growth and also in August, stem yield was smaller when row width was larger; in September row width did not affect stem yield or quality (van der Werf et al.,1995).

THE CHEMICAL COMPOSITION OF HEMP STEM

Hemp stems can be separated into two components: the stem tissues outside the vascular cambium (bark) and the stem tissues inside the vascular cambium (core). Bark and core differ in their chemical composition: Bedetti and Ciaralli (1976) reported 67 percent cellulose, 13 percent hemicellulose and 4 percent lignin in the bark of an Italian hemp cultivar. Its core contained 38 percent cellulose, 31 percent hemicellulose, and 18 percent lignin.

The bast contains primary and secondary fibers, which are thick-walled, have a high cellulose and low hemicellulose and lignin content. Primary bast fibers are 5 to 40 mm long and heterogeneous, secondary bast fibers are smaller and uniformly about 2 mm long. The proportion of bast fibers in the stem and the ratio primary/secondary bast fibers depends on cultivar and development stage (age) of the plant.

The woody core contains parenchyma, vessels (both having transport functions in the plant) and libriform fibers, which will be referred to as core fibers, for rigidity and strength. Core fibers are thin-walled and short (0.55 mm) and have a chemical composition that resembles hardwood; approximately 40 percent cellulose, 20 percent hemicellulose, and 20 percent lignin.

Fiber length and the contents of cellulose and lignin are important quality parameters of raw materials for paper. The strength of paper increases with fiber length. Cellulose content is important, because in chemical pulping the pulp yield corresponds to the cellulose content of raw material. Since lignin is removed by environmentally unfriendly procedures, a low lignin content is desirable. In the bark, fibers are longer, cellulose content is higher, and lignin content is lower than in the core. As a result, bark is more valuable as a raw material for paper than core and the value of hemp stems therefore depends primarily on its bark content (van der Werf et al., 1995).

Contrary to wood, with the average composition of many annual rings, in hemp the ratio bast/core, the chemical composition and the core fiber morphology change during the growing season. The differences in fiber morphology and chemical composition and the seasonal effects make hemp a nonuniform raw material, which affects both the types of pulps that can be produced from it and the required technology to produce these pulps.

The effect of development stage of the plant on strength properties of hemp bast fibers was studied in view of their application in paper. Both thermo mechanically produced handsheets from bast fibers and native bast fibers of hemp harvested at three different dates were evaluated. Handsheet properties varied some 10 percent but were not significantly different over the period from August until October. Direct zero-gauge tenacity measurements showed no significant differences between unprocessed fibers of different harvest dates, and between primary and secondary bast fibers. The $1/8''$-gauge tenacity of primary and secondary bast fibers differ significantly (Riddlestone and Franck, 1995).

CULTIVATION TECHNIQUES AND CROP DESTINATION

The harvesting and subsequent processing techniques of the raw material depends on the purpose of the crop: textile use or paper pulp use.

Hemp for Textile Use

The centuries-old method of hemp textile production involves:

1. Harvesting after flowering but before the seeds set, when the stems are whitening at the base and the leaves are starting to drop. The fiber content is reduced and becomes coarser toward seed formation. Where it is desired to obtain fiber and seed, the male plants are first collected by hand pulling, and the female plants are left to enable the seeds to ripen.
2. "Retting" the crop. Retting is the name given to the process whereby bacteria and fungi break down the pectins that bind the fibers to the stem allowing fiber to be released; one of two alternative methods was generally used:

   - *Water retting* involves laying the stems in water in tanks, ponds, or in streams for about ten days; it is more effective if the water is warm and bacteria laden. Fiber extraction from fiber crops by traditional retting methods is highly polluting or carries high risks of crop failure and yields of varying fiber qualities over the years. Moreover, top quality fibers are not constantly available and raw material price fluctuations are of concern. Alternatively, nonpolluting processing techniques, which guarantee constant fiber qualities for industrial buyers, are urgently needed. Water retting is unlikely to be viable on a modern farm as it is awkward and time-consuming and produces an effluent that can be a source of pollution.
   - *Dew retting* entails laying the crop on the ground for ten to thirty days, turning as necessary to allow even retting. After cutting, the hemp stems were laid parallel in rows to dew ret. The earlier the crop was harvested, the faster retting was completed. The stems needed turning at least once (some-

times twice) in order to allow for even retting. When turning, it could happen that the stems close to the ground remained green while the top was retting and turning brown. When retting is complete the crop is entirely brown/grey. The thicker stems took longer to ret. Therefore uniformly tall, fine stems would seem to be the best for trouble-free retting.

Judging the degree and completeness of retting is a subjective exercise based upon experience: retting seems to be complete when the fiber bundles appear white, separate from the woody core, and divide easily into individual, finer fibers for their full length.

Once it is considered that retting has gone far enough, the crop needs to be dried to halt the retting process before it damages the fiber and to prevent further retting in storage (a moisture content of less than 16 percent is recommended).

3. Breaking the stems by passing through a "breaker" or fluted rollers.
4. Separating the fiber from the woody core ("scutching") by beating the broken stems with a beech stick or passing through rotary blades.
5. "Hanckling" (combing) to remove any remaining woody particles and to further align the fibers into a continuous "sliver" for spinning.

Yield

The yield of the hemp is between 7 and 10 tonnes (metric tons) dm/ha and the fiber yield will be approximately 1.5 to 2.0 ton/ha. The amount of fiber contained within the stem is about 30 percent—of which perhaps 20 percent is suitable for textiles. The plant consists approximately of 10 percent roots, 60 to 70 percent stem, 15 to 20 percent leaves and 5 to 15 percent seed.

Quality Criteria

The quality of hemp is judged mainly by its color and luster; high-quality fiber should be lustrous and give a decided snap when broken. Hemp fiber is longer than flax fiber, but is less flexible and more coarse. It does not bleach well, and as it lacks elasticity and flexibility it is not used for fine textiles. The ultimate fiber cells vary in length from 5 to 55 mm, and have an average length of about 20 mm, and their diameter varies between 0.016 and 0.050 mm.

Quality criteria for hemp fibers are bast fiber content, proportion of secondary fiber in the bast fraction, and xylem fiber length. For hemp, content of xylem, primary bast fiber, secondary bast fiber, xylem (libriform) fiber length and diameter are important.

There are several quality criteria for fibers and they can be divided into general criteria and application-related criteria. The general criteria are the dimensions of the fibers (length-diameter ratio), density, brittleness, elasticity, tensile strength, strain modulus, wear resistance, purity, fineness, absorbing capacity, thermal and fire resistance, low variations in properties, cleanness, degradability, and durability. The most important process-related criteria are exchangeability of materials, processing speed, processing costs, and amount of waste production.

Logistics

The logistic chain for plant fiber, from crop to end use, as it is now, is very long and complicated. Beginning with farmers and breeders, the chain is followed by traders, scutchers, hacklers, various industries to the end-use market, and customers. The chain from farmer to industry is too long, resulting in problems with supply and high costs. A logistic support system is required. The low density of plant fibers causes high costs for transportation and storage. Climatological variations through the years result in differing quantities produced. Constant quantity supply can be achieved when the logistics of the chain are controlled; e.g., with quota, contract cultivation, spreading of cultivation over different EU countries, by keeping buffers, or by import of fibers from outside the EU. The need for constant quantity supply can be minimized when plant fibers are exchangeable in the production process (van Dam, 1995).

Two Problems Need to Be Overcome

First, as hemp is harvested late in the season (a month later than flax) dew retting of hemp is unreliable. Therefore it is necessary to develop retting technologies that are suitable for the temperate climate, or bypass the need to depend on the weather. A lot of work has been done on retting, especially on flax, particularly by the French. A great deal of work still needs to be done. What is certain is that unless the problem of retting is overcome it will not be possible to produce textiles from hemp economically in countries with temperate climates.

The second problem is technically easier to overcome, but still needs substantial research and development. After satisfactorily retting the hemp stems, the fiber needs to be removed from the rest of the plant. The only machinery that exists, to our knowledge, is adapted flax machinery that is not entirely suited to hemp, which requires a larger and more robust machine.

Hemp for Paper Pulp

Fiber hemp may be an alternative to wood as a raw material for pulp and paper production (Meijer et al., 1995). The economic feasibility of hemp as a raw material for paper largely depends on its yield. Reports on the yields obtained in Europe are poor and refer to straw, i.e., field-dried stems containing 10 to 18 percent moisture and remnants of the inflorescence and leaves. For Denmark, Nordestgaard (1976) reported straw yields of 9 t ha^{-1}; in Poland, Jaranowska (1964) obtained 7 to 9 t ha^{-1}; Mathieu (1980) reported 8 to 10 t ha^{-1} in France. In Italy higher straw yields are obtained, e.g., 15 t ha^{-1} (Marras and Spanu, 1979); in the Netherlands, Aukema and Friederich (1957) obtained 10 to 13 t ha^{-1} and van der Schaaf (1966) 12 t ha^{-1} of straw. The seeding rate appears to be a major factor determining yield and quality of fiber hemp, however its effect on stem yield and quality varies.

Both bast and core have to be processed and the objective is to cultivate a crop with a maximum fiber yield. In this crop two harvesting systems are used, depending on whether the seed is to be harvested or not (FNPC, 1985): (1) for the best and highest fiber yield, the hemp is cut with a mower conditioner for field drying; in about four days, the moisture content is under 15 percent and the hemp is baled with a round baler. With this moisture content the material can be stored for a long time; (2) if the seed also has to be harvested, the stem tops are cut later and threshed with a combine (in the first week of September, in France); subsequently, the crop is mown and laid down in a swath to dry. Tedding the swath will increase drying speed and quality evenness. With this second harvesting system the fiber yield is much lower (depending on the cutting height of the combine) and because of poor weather conditions in September, the fiber quality can be lower.

As is known, storage with conservation of quality is possible when the moisture content of the material is below 15 percent. Drying in September in temperate countries (e.g., the Netherlands) is expected to be too much of a weather risk, resulting in nonconstant and nonpredictable quality of the raw material. Therefore, a system with conservation by making silage was developed, called *wet harvesting method*. This conservation is attractive for pulp making because less energy is needed for pulping and separating bast and core is easy. A system for preservation and storage is also needed to supply a pulp processing plant for a whole year.

Finally, the processing of hemp into paper pulp demands a separation of bark and core because these fractions differ physically (fiber length and diameter) and chemically (cellulose and lignin content). The separation may be done as a harvest operation in the field or as a preprocessing step of the pulping.

Chopping of the standing crop is less weather dependent and after chopping, the hemp has a dm content of 33 percent. Artificial drying is too expensive and alternatively an attempt to preserve the hemp by ensiling was tried (wet anaerobic preservation). In corn or grass silages lactic acid bacteria consume available sugars and produce lactic acid up to a pH of 4.2. The product is then preserved and no further losses occur (McDonald, Henderson, and Heron, 1991). In hemp silage the pH did not drop to 4.2, not even after addition of lactic acid bacteria. The reason for this was the lack of the necessary sugars. Per kg stem dry matter only 4 g of sugars were available and based upon grass silage, about 75 g are needed for a sufficient fermentation (Wieringa and de Haan, 1961). Despite the inadequate fermentation, no cellulose or some hemicellulose was lost in silage.

Separation of Bark and Core

Three methods can be used to separate bark and core: decortication of green stems in the field, separation of bark and core after chopping and ensiling, and by sieving or flotation. With decortication the bark cannot be cleaned well enough as it still contains more than 40 percent of core. The quantity of fixed core mainly depends on stem diameter: the smaller the diameter, the higher the quantity of fixed core. After chopping, about 90 percent of the mixture consists of clean bark and core pieces and this makes it possible to separate bark and core by sieving or flotation. Sieving of chopped hemp results in a "pollution" of both bark and core of about 25 percent. Thus the size difference between the chopped bark and core is not big enough to separate them by sieving. With flotation, the hemp is thrown in a bath with slow-moving water in which some of it floats and the rest sinks. Very clean bast can be collected with this operation.

Conclusively, decortication is less effective and more expensive, while wet anaerobic preservation and separation of bark and core by flotation is the most uncertain method for the supply of consistent high-quality fibers throughout the year. Acidic and alkaline preservation techniques seem to be the best methods, but more research is needed on these processes, with special attention paid to the observed decrease in fiber strength (de Maeyer and Huisman, 1994).

CONSTRAINTS TO DRY MATTER PRODUCTION IN FIBER HEMP

Research aimed at assessing the potential productivity of fiber hemp and to identify constraints to that productivity was conducted. Growth analyses were performed on hemp crops in three consecutive years, using several cultivars and seeding rates (Meijer et al., 1995).

The number of living plants m^{-2} ranged from 86 to 823 at emergence, depending on treatment, and from 38 to 102 at final harvest. Increased seeding rates led to earlier canopy closure and higher initial biomass production, but more plants died during the growing season and the stem yield at final harvest was not affected by seeding rate. Average radiation-use efficiency (RUE); aboveground accumulated dry matter divided by intercepted photosynthetically active radiation) for the entire growing season under favorable growing conditions was 1.9 gMJ^{-1}. These values are rather low, relative to the RUEs prior to flowering of 2.2 to 2.9 reported for other C3 species such as sunflower, rice, wheat, potato, and chicory (Kiniry et al., 1989; Haverkort et al., 1992; Meijer, Mathijssen, and Borm, 1993).

A low RUE is expected when the biomass contains much fat, protein, or lignin. The costs and weight losses associated with the conversion of primary assimilates to those constituents are high compared to these associated with the synthesis of starch or cellulose (Penning de Vries, Brunsting, and van Laar, 1974). Hemp stems contain about 15 percent lignin (Bedetti and Ciaralli, 1976), whereas the storage organs of carbohydrate-producing crops such as wheat and sugar beet contain 5 to 6 percent lignin (Vertregt and Penning de Vries, 1987). Because of the relatively high lignin content of its stems, hemp is expected to form about 8 percent less stem dry matter from the same amount of primary assimilates than wheat, potato, or sugar beet produce in their storage organs. This corresponds to a 6 to 7 percent lower aboveground dry matter yield and a similar reduction in RUE.

The formation of about 1 tha^{-1} of seed containing 40 percent oil and almost 30 percent protein (Sinclair and de Wit, 1975) reduces RUE after flowering. The primary assimilates required for 1 t of hemp seed could have yielded 1.4 t of dry matter in vegetative plants, thereby increasing aboveground dry matter production by 2.5 percent. The large amount of assimilate required to synthesize oil and protein accounts in part for the low RUE during the final growth phase.

In conclusion, the relatively low dry matter production of fiber hemp per unit of intercepted radiation is the result of several factors. The rate of canopy photosynthesis is negatively affected by the high extinction coefficient of the hemp canopy. Dry matter production is reduced by 6 to 7 percent as a result of conversion losses during the synthesis of a relatively large quantity of lignin in the stem. After flowering, dry matter production is reduced by the synthesis of fat and protein in the seed and the senescence of the canopy.

IMPLICATIONS FOR FUTURE RESEARCH

This analysis indicates avenues that might lead to greater productivity of hemp:

1. Earlier canopy closure may improve crop productivity by increasing the total amount of intercepted radiation during the growing season. Earlier sowing may be a more effective way of obtaining earlier canopy closure than high seeding densities. The effect of such practices remains to be investigated.

2. Conversion losses may be reduced by breeding hemp varieties with a lower lignin content in the stem; this would also improve the quality of stem dry matter as a raw material for pulp and paper.
3. Using cultivars that flower later than the current ones would probably improve stem yield, as no assimilates would be invested in the inflorescence, and new leaves would continue to be formed, thereby prolonging the maximum photosynthetic capacity of the canopy.
4. In the cultivation phase, further experimental investigations are particularly needed, to a greater or lesser degree, in the sectors of fertilizers, of plant density (in relation to the variety planted, the cultivation environment, fertilizers used, etc.), of the best time to harvest the crop (which has a bearing on the productivity and the quality of the fiber), of the fight against disease and insect pests, of cultivation machinery, and particularly of mechanization of the operations needed in country retting. For several reasons, the future of the hemp industry must be oriented toward industrial retting. The trend must be toward research and experiments directed toward finding a solution for the basic problems of the economics of the operation and the quality of the product. In particular: (a) preparation of the raw material on the farm, so as to reduce it to the smallest possible volume (which involves perfecting the operation of hatchelling on green hemp); (b) perfection of the process of retting, through biological, biochemical, or chemical means; (c) reduction to a minimum of the time required for retting and subsequent operations.
5. Various fiber separation processes, based on the use of steam explosion, detergents, or ultrasound have now been developed on a laboratory scale. Although the quality of the produced fibers has been appreciated with keen interest by industry, these technologies still lack demonstration on the pilot and production scales. Most advanced at this point are purely mechanical methods that combine harvesting and separation. They provide a limited range of product qualities or serve as a preprocessing step for the already mentioned biochemical or physical processes.

REFERENCES

Allavena, D. (1964). "Aspetti e problemi del miglioramento genetico della canapa monoica." *Sementi Elette*, 1:20-26.
Aukema, J. J. and Friederich, J. C. (1957). "Report on the experiments with hemp in the years 1952-1956. Dutch Flax Institute." *Wageningen, Report 33*, pp. 25.
Baraniecki, P., Grabowska, L., and Mankowski, J. (1995). "Recultivation of degraded areas through cultivation of hemp." *Book Abstracts of Symposium on Hemp*, Frankfurt, March 2-5, 1995.
Bedetti, R. and Ciaralli, N. (1976). "Variazione del contenuto di cellulosa durante il periodo vegetativo della canapa. [Aspects of the cellulose content during the vegetative period of hemp.]" *Cellulosa e Carta*, 26:27-30.
Bredemann, G., Schwanitz, F., and von Sengbusch, R. (1957). "Auhgaben und Moglichkeiten der modernen Hanfzuchtung mit besonderer Berucksichtigung des Problems der Zuchtung haschischarmer oder-freier Hanfsorten." *Max Plank Institut für Zuchtungsforschung, Hanburg-Volksdorf, Technical Bulletin* 4:1-15.
de Maeyer, E. A. A. and Huisman, W. (1994). "New technology to harvest and store fiber hemp for paper pulp." *Journal of the International Hemp Association*, 1(2):38-41.
Faeti, V., Mandolino, G., and Ranalli, P. (1996). "Genetic diversity of *Cannabis sativa* L. germplasma based on RAPD markers." *Plant Breeding*, 115:367-370.
Fleischmann, R. (1938). "Der Einfluß der Tageslange auf der Entwick lungsrhythmus von Hanf und Ramie." *Faserforschung*, 13:93-99.
FNPC (1985). "Le chanvre monoique: La récolte." *Fédération Nationale des Producteurs de Chanvre*, Le Mans, France: p. 11.
Grassi, G., Moschella, A., and Fiorilli, M. (1997). "Evaluation and development of serological methods to detect THC in hemp." *Book Abstracts of Symposium on Hemp*, Frankfurt, February 27-March 2, 1997.
Haverkort, A. J., Boerma, M., Velema, R., and van de Waart, M. (1992). "The influence of drought and cyst nematodes on potato growth. 4. Effects on crop growth under field conditions of four cultivars differing in tolerance." *Netherland Journal of Plant Pathology*, 98:179-191.
Hoffmann, W. (1961). "Hanf, *Cannabis sativa*." In *Handbuch der Pflanzenzuchtung*, H., Kapport and W. Rudorf (Eds.), pp. 204-261. Band V. Berlin: Paul Parey.
Huhnke, W., Jordan, Ch., Neuer, H., and von Sengbusch, R. (1951). "Grundlagen fur die Zuchtung cines. Monozischen Hanfes." *Z Pflanzenzucht*, 29:55-75.
Jakobey, I. (1965). "Experiments to produce hemp with fine fiber." *Novenytermeles*, 14:45-54.
Jaranowska, B. (1964). "Effect of increasing doses of nitrogen and seeding rate on monoeciuos and dioecious hemp." *Yearbook* Inst., Przem. Wlok. Lyk., Poznan, Poland.

Kiniry, J. R., Jones, C. A., O'Toole, J. C., Blanchet, R., Cabelguenne, M., and Spanel, D. A. (1989). "Radiation-use efficiency in biomass accumulation prior to grain-filling for five grain-crop species." *Field Crop Research*, 20:51-64.

Kok, C. J., Coenen, G. C. M., and de Heij, A. (1994). "The effect of fiber hemp (*Cannabis sativa* L.) on soil-borne pathogens." *Journal of the International Hemp Association*, 1:6-9.

Lotz, L. A. P., Groeneveld, R. M. W., Habekotte, B., and van Oene, H. (1991). "Reduction of growth and reproduction of *Cyperus esculentus* by specific crops." *Weed Research*, 31:153-160.

Mandolino, G., Faeti, V., Carboni, A., and Ranalli, P. (1997). "A 400 RAPD marker tightly linked to the male phenotype in dioeciuos hemp." *Book Abstracts of Symposium on Hemp*, Frankfurt, February 27-March 2, 1997.

Marras, G. F. and Spanu, A. (1979). "Aspetti di tecnica colturale in canapa da cellulosa. Densità di semina e concimazione azotata. [Aspects of cultural practices in hemp for cellulose. Seeding density and nitrogen consumption.]" *Annali Facoltà Agraria Università Sassari*, XXVII.

Mathieu, J. P. (1980). "Chanvre." *Technical Agriculture*, 5:1-10.

McDonald, P., Henderson, N., and Heron, S. (1991). *The biochemistry of silage*, second edition, p. 340. Chalcombe Publications, Marlow, England.

Meijer, W. J. M., Mathijssen, E. W. J. M., and Borm, G. E. L. (1993). "Crop characteristics and inulin production of Jerusalem artichoke and chicory." In *Inulin and inulin-containing crops: Studies in plant science. 3*, A. Fuchs (Ed.), pp. 29-38. Amsterdam: Elsevier.

Meijer, W. J. M., van der Werf, H. M. G., Mathijssen, E. W. J. M., and van den Brink, P. W. M. (1995). "Constraints to dry matter production in fiber hemp (*Cannabis sativa* L.)." *European Journal of Agronomy*, 4(1):109-117.

Meijer de, E. P. M. and Keizer, L. C. P. (1994). "Variation of *Cannabis* for phenological development and stem elongation in relation to stem production." *Field Crop Research*, 38:37-46.

Meijer de, E. P. M. (1995). "Fiber hemp cultivars: A survey of origin, ancestry, availability and brief agronomic characteristics." *Journal of the International Hemp Association*, 2(2):66-73.

Monteith, J. L. (1977). "Climate and the efficiency of crop production in Britain." *Philosophiae Transactions of the Royal Society, London*, 281:277-294.

Nordestgaard, A. (1976). "Varietal experiments and experiments on fertilization and premature ripening of spinning hemp (*Cannabis sativa* L.)." *Report 1325 of the State Research Station, Roskilde*.

Penning de Vries, F. W. T., Brunsting, A. H. M., and van Laar, H. H. (1974). "Products, requirements, and efficiency of biosynthesis: A quantitative approach." *Journal of Theoretical Biology*, 45:339-377.

Ranalli, P., Di Candilo, M., Marino, A., Grassi, G., Polsinelli, M., and Casarini, B. (1997). "Induction of morphological mutants in *Cannabis sativa* L." *Book Abstracts of Symposium on Hemp*, Frankfurt, February 27-March 2, 1997.

Ranalli, P., Di Candilo, M., Marino, A., Zottini, M., Fuochi, P., Polsinelli, M., and Casarini, B. (1996). "Induzione di mutanti in *Cannabis sativa* L." *Sementi Elette*, 2:49-55.
Riddlestone, S. and Franck, B. (1995). "Hemp for textiles." *Book Abstracts of Symposium on Hemp*, Frankfurt, March 2-5, 1995.
Sinclair, T. R. and de Wit, C. T. (1975). "Photosynthate and nitrogen requirements for seed production by various crops." *Science*, 189:565-567.
Spitters, C. J. T. (1990). "Crop growth models: Their usefulness and limitations." *Acta Horticulturae*, 267:349-368.
Tamm, E. (1933). "Further investigations on the germination and the emergence of agricultural crop plants." *Pflanzenbau*, 10:297-313.
Tarasov, A. V. (1975). "Hemp yield and yield of other crops in a technical rotation at different growing systems" (in Russian). In *Biology, cultivation, and the primary processing of hemp and kenaf*. All Union Scientific and Research Institute of Bast Crops, Glukhov 38:83-88.
van Dam, J. E. G. (1995). "Potentials of hemp as industrial fiber crop." *Book Abstracts of Symposium on Hemp*, Frankfurt, March 2-5, 1995.
van der Schaaf, A. (1966). "Quantitative and qualitative influence of some cultural practices on hemp (*Cannabis sativa* L.)." *Fibra*, 11:1-8.
van der Werf, H. M. G. (1997). "The effect of plant density on development and light interception in hemp (*Cannabis sativa* L.)." *Book Abstracts of Symposium on Hemp*, Frankfurt, February 27-March 2, 1997.
van der Werf, H. M. G., Brouwer, K., Wijlhuizen, M., and Withagen, J. C. M. (1995). "The effect of temperature on leaf appearance and canopy establishment in fiber hemp (*Cannabis sativa* L.)." *Annals of Applied Biology*, 126:551-561.
van der Werf, H. M. G., Haasken, H. J., and Wijlhuizen, M. (1994). "The effect of daylength on yield and quality of fiber hemp (*Cannabis sativa* L.)." *European Journal of Agronomy*, 3:117-123.
van der Werf, H. M. G., van Geel, W. C. A., and Wijlhuizen, M. (1995). "Agronomic research on hemp (*Cannabis sativa* L.) in the Netherlands, 1987-1993." *Book Abstracts of Symposium on Hemp*, Frankfurt, March 2-5, 1995.
van der Werf, H. M. G., van Geel, W. C. A., van Gils, L. J. C., and Haverkort, A. J. (1995). "Nitrogen fertilization and row width affect self-thinning and productivity of fiber hemp (*Cannabis sativa* L.)." *Field Crop Research*, 42:27-37.
Vertregt, N. and Penning de Vries, F. W. T. (1987). "A rapid method for determining the efficiency of biosynthesis of plant biomass." *Journal of Theoretical Biology*, 128:109-119.
Wieringa, G. W. and de Haan, S. (1961). *Inkuilen* (Ensiling), IBVL, Wageningen, p. 49.
Zottini, M., Mandolino, G., and Ranalli, P. (1997). "Effects of γ-ray treatment on *Cannabis sativa* pollen viability." *Plant Cell, Tissue, and Organ Culture*, 47:189-194.

Chapter 5

Crop Physiology of *Cannabis sativa* L.: A Simulation Study of Potential Yield of Hemp in Northwest Europe

Hayo M. G. van der Werf
Els W. J. M. Mathijssen
Anton J. Haverkort

INTRODUCTION

Hemp (*Cannabis sativa* L.) is grown for the production of fiber (Rabelais, 1546), cannabinoids (Beaudelaire, 1860), or seed (Deferne and Pate, 1996). *Cannabis* originates from Central Asia but has been cultivated from the Equator to the Polar Circle (Vavilov, 1926). Human use of hemp goes back at least 6,000 years and it may be one of the oldest nonfood crops (Schultes, 1970). For thousands of years, hemp bast fiber has been used to manufacture rope, fabric, and paper. Cannabinoids have been used for medical, spiritual, and recreational purposes, whereas the seed was produced mainly for its oil.

From the sixteenth to the eighteenth century, hemp and flax (*Linum usitatissimum* L.) were the major fiber crops in Russia, Europe, and North America (Pounds, 1979; Abel, 1980). Both crops were used for the production of fabrics for garments. Worn-out flax and hemp fabrics were used as raw materials in paper mills. How-

---

This chapter was originally published as a paper in *Annals of Applied Biology*, Volume 129: 109-123. Copyright © 1996 by The Association of Applied Biologists. Reprinted with permission.

Our thanks to W. J. M. Meijer and P. C. Struik for their helpful comments on the manuscript.

ever, the large-scale cultivation of cotton, jute, and other tropical fibers, and the development of new technologies to process wood into paper pulp caused the world area of hemp and flax to decline in the nineteenth century. This decline has continued in this century, due to the advent of synthetic fibers. The presence of psychoactive components in hemp was another reason for its decline, as this became a motive to prohibit hemp cultivation in many countries (Dempsey, 1975). Since World War II, the main areas of fiber hemp production have been in China, the Soviet Union, and Eastern Europe. In 1994 there were 119,000 ha under fiber hemp worldwide (FAO, 1995).

During recent decades the paper pulp industry has been criticized for its negative impact on the natural environment through deforestation, or the replacement of old-growth forests by tree plantations (Postel and Ryan, 1991), the emission of chemical waste, high energy use, and the production of toxic and mutagenic waste products by chlorine bleaching (McDougall et al., 1993). Measures taken to tackle these problems include increased recycling of paper, more sustainable management of tree plantations and forests, and a shift toward less harmful pulp and paper technologies.

Intensive cotton (*Gossypium* L.) production has also been severely criticized for its negative effects on the environment through intensive use of pesticides (cotton can be treated twenty times per season), and high fertilizer and irrigation requirements (Pesticides Trust, 1990; Pimentel et al., 1991). These problems can be reduced to some extent by introducing integrated pest management techniques or by shifting to organic farming methods (Pimentel et al., 1991; Pleydell-Bouverie, 1994).

A comeback of hemp as a raw material for paper and textile may further contribute to the sustainability of the paper and textile industry. Growing an annual crop on farmland to produce fiber obviously lessens the need to cut down forests. In addition, less energy is required to produce pulp from hemp than from wood (van Roekel et al., 1995), and the lignin content of the former is lower, offering better opportunities for nonchlorine bleaching or the production of unbleached pulp (McDougall et al., 1993). Relative to cotton, hemp can be produced more sustainably, as it requires little pesticide and its fertilizer requirements are modest.

From World War II until the 1980s, hemp was a largely forgotten crop. However, in eastern and central Europe and in France breeding work continued (de Meijer, 1995), leading to more productive hybrid varieties (Bócsa, 1971), increased fiber contents (Bócsa, 1995) and very low contents of psychoactive substances (Fournier et al., 1987; Goloborod'ko, 1995). The potential of hemp as an attractive crop for sustainable fiber production was pointed out in the early 1980s (Hanson, 1980). Its yield was reported to be high, and it was said to improve soil structure (Du Bois, 1982). Furthermore, hemp was claimed to suppress weeds effectively, and to be virtually free from diseases or pests. A few years later, Herer (1985) claimed that hemp yielded several times more cellulose than other crops such as corn (*Zea mays* L.), kenaf (*Hibiscus cannabinus* L.), or sugar cane (*Saccharum* L.).

As a result of this renewed interest in hemp, preliminary research was conducted during the 1980s into the best ways of growing, harvesting, and pulping fiber hemp in the Netherlands (e.g., du Bois, 1984; de Groot, van Zuilichem, and van der Zwam, 1988). The results were encouraging, and in January 1990 a comprehensive four-year study, the Hemp Research Programme, was started to investigate the potential of fiber hemp as a new raw material for the pulp and paper industry, and to establish whether the production of fiber hemp for paper pulp would be economically attractive. The major research disciplines within the program were: plant breeding (de Meijer, 1994; Hennink, 1994); crop physiology and agronomy (van der Werf, 1994); plant pathology (Kok, Coenen, and de Heij, 1994); harvest and storage technology (de Maeyer and Huisman, 1994); pulp technology (de Groot et al., 1994; van Roekel, 1994); and economics and market research.

From this research program, it was concluded that fiber hemp has potential as a profitable crop for arable farmers in the Netherlands, provided a pulp factory is set up (Bakker and van Kemenade, 1993). Agronomically, hemp proved to be attractive, as most of the claims made by early hemp advocates proved to be true (van der Werf, van Geel, and Wijlhuizen, 1995): hemp can supply high fiber yields, requires little or no pesticide, and suppresses weeds and some major soil-borne diseases. However, in the maritime climate of the Netherlands the crop is not disease-free, as the fungus *Botrytis cinerea* can cause severe damage in wet years (van der Werf, van Geel, and Wijlhuizen, 1995). In spite of this, hemp will manifestly fit into sustainable farming systems.

This chapter examines the crop physiological and agronomic characteristics of hemp, and a simple crop growth model is used to assess its yield potential. Hemp is compared with kenaf, another annual fiber crop. This study is based on recent research conducted in the Netherlands and on data from the literature.

CROP PHYSIOLOGICAL CHARACTERISTICS

A crop achieves its full potential when not limited by shortage of water or of nutrients, by pest or disease attack, or by other stresses. In such ideal conditions, its dry matter production is approximately proportional to the amount of light (photosynthetically active radiation [PAR]) intercepted by the crop canopy (Monteith, 1977). Dry matter yield (Y) of such a nonstressed crop can be described as: $Y = L \times RUE \times HI$ where L is the amount of light intercepted during a growing season, RUE is the radiation use efficiency (the amount of dry matter produced per unit of light intercepted), and HI is the harvest index (the proportion of total dry matter consisting of plant parts of economic value). These three parameters will be considered.

Interception of Light During the Growing Season

The amount of light an annual crop intercepts depends on its emergence date, its rate of canopy establishment, the proportion of incident light intercepted by a fully established canopy, the date of onset of canopy senescence, and the rate at which light interception by the canopy declines during senescence. These factors can be affected by environmental parameters (temperature, radiation, daylength) and by crop management. The main crop management decisions affecting light interception by a nonstressed fiber hemp crop are cultivar, plant density, sowing date, and harvest date. To avoid cultivar and plant density limiting hemp yield, the cultivar should not flower (van der Werf, Haasken, and Wijlhuizen, 1994) and plant density should be sufficiently high without exceeding the maximum density that can be sustained at the expected yield (van der Werf, Wijlhuizen, and de Schutter, 1995; van der Werf and van den Berg, 1995). We will examine the effect of sowing date and harvest date on light interception by a nonflowering hemp crop grown at a plant density appropriate for a high yield.

Sowing Date

In northwest Europe, incoming radiation is greatest in May, June, and July, whereas temperatures are highest in July and August (see Table 5.1). In May and June, the interception of incident light by spring-sown crops is generally incomplete, because of slow canopy expansion at suboptimum temperatures. However, hemp grows at low temperatures, and might therefore be well adapted to the temperate climate of northwest Europe. Its base temperature for leaf appearance is 1° C, and for canopy establishment is 2.5° C (van der Werf, Brouwer et al., 1995). In this respect, hemp is similar to one of the major arable crops in northwest Europe, namely sugar beet, also a spring-sown dicotyledon. Sugar beet has a base temperature of 1° C for leaf appearance and of 3° C for leaf expansion (Milford, Pocock, and Riley, 1985). From sowing to 50 percent plant emergence, sugar beet requires about 90° Cd (base 3° C) (Smit, 1989), whereas hemp requires 56° Cd (base 3° C) (van der Werf, Brouwer et al., 1995). To reach canopy closure, sugar beet grown at its optimal density of about eight plants m^{-2} requires another 500° Cd (base 3° C) (Milford et al., 1985), whereas hemp (at sixty-four plants m^{-2}) requires another 340° Cd (base 2.5° C) (van der Werf, Brouwer et al., 1995). As a result, under similar circumstances, hemp establishes a closed canopy more rapidly than sugar beet.

Table 5.1. Average global radiation and temperature at De Bilt, 1961-1990.

Parameter	MONTH							
	March	April	May	June	July	August	Sept.	Oct.
Mean temperature (° C)	5.0	8.0	12.3	15.2	16.8	16.7	14.0	10.5
Global radiation (MJ m^{-2} d^{-1})	7.9	12.9	16.8	17.9	16.7	14.7	10.3	6.1

Source: Summarized from data supplied by the Dutch Royal Meteorological Institute.

To maximize the yield of sugar beet in the Netherlands, the crop should be sown from the end of March, as soon as soil and weather conditions permit (Smit, 1993). As sugar beet and hemp have similar base temperatures for growth, one might expect their optimal sowing dates to be similar. However, frost resistance is another factor that may affect the optimal sowing date of a crop. Sugar beet is most sensitive to frost at emergence; seedlings may be killed by a frost of about $-5°$ C, though once fully emerged, the plants tolerate frosts of up to $-10°$ C (A. L. Smit, personal communication, 1994). Hemp seedlings survive a short frost of up to $-8°$ to $-10°$ C (Grenikov and Tollochko, 1953); older hemp plants tolerate frosts of up to $-5°$ to $-6°$ C (Senchenko and Timonin, 1978). Sugar beet is more at risk from frost during emergence than hemp, whereas hemp is more at risk than sugar beet for a much longer period.

To estimate the effect of sowing date on canopy establishment and potential light interception by fiber hemp in the Netherlands, a simple crop growth model was used, based mainly on the results of van der Werf, Brouwer et al. (1995). The model is a modified version of the "light interception and utilization" (LINTUL) model proposed by Spitters (1990). It was assumed that the crop required $77°$ Cd (base $1°$ C) from sowing to emergence (van der Werf, Brouwer et al., 1995). Canopy establishment was described using the relation between the proportion of light intercepted and thermal time for crops grown at sixty-four plants m^{-2} (van der Werf, Brouwer et al., 1995). Maximum interception was assumed to be 99 percent until harvest (van der Werf, Brouwer et al., 1995; van der Werf, Wijlhuizen et al., 1995). Average (1961 to 1990) temperature and radiation data recorded at De Bilt in the center of the Netherlands were input in the model.

The second half of April is often recommended as the best period for sowing hemp (Heuser, 1927; de Jonge, 1944; Senchenko and Demkin, 1972; Mathieu, 1980). For a hemp crop sown on April 15, the model calculated emergence on April 26 and canopy closure (90 percent PAR interception) on June 1 (see Table 5.2). For a crop sown thirty days earlier, on March 16, emergence and canopy establishment would take longer, but canopy closure would still be advanced by twelve days and intercepted PAR would increase by 120 MJ m^{-2}. Sowing hemp on May 15 instead of April 15 would delay canopy closure by nineteen days and intercepted PAR would decrease by 185 MJ m^{-2} (see Table 5.2).

Table 5.2. Simulated effects of sowing date on the date of emergence and canopy establishment and on the accumulated intercepted PAR until August 1 by the canopy of a hemp crop grown at a plant density of 64 m^{-2}. Average (1961-1990) temperature and radiation data were input in the model.

Sowing date	Sowing to emergence (days)	Emergence to 90% interception (days)	Date of 90% interception	PAR intercepted until August 1 (MJ m^{-2})
March 16	16	49	May 20	737
March 31	13	42	May 25	686
April 15	11	36	June 1	617
April 30	8	32	June 9	538
May 15	7	29	June 20	432

In conclusion, therefore, as hemp grows well at low temperatures, advancing its sowing date from April 15 to March 16 or 31 will advance canopy closure and increase the amount of PAR intercepted by the canopy. However, advancing the sowing date will also increase the probability of frost damage. This risk should be taken into account, particularly at frost-prone sites.

Harvest Date

The currently available French and Hungarian cultivars flower in August, and after flowering stem growth slows down and ceases in the first half of September (Meijer, van der Werf, Mathijssen et al., 1995; van der Werf, Haasken, and Wijlhuizen, 1994; van der Werf, Wijlhuizen, and de Schutter, 1995; van der Werf, van Geel et al., 1995). To obtain maximum stem yield these cultivars should be harvested in early September. In later-flowering cultivars stem growth continues longer, and optimum harvest date will be later.

Traditionally, harvesting hemp involves a period of field drying. In the Netherlands, the weather in September is rarely favorable for field drying of the crop, and for this reason, the potential of ensiling as an alternative way of preserving hemp stems was investigated (de Maeyer and Huisman, 1994). The results obtained so far indicate that ensiling is a promising, but more expensive, technique than field drying. Field drying involves harvesting in August and, as a result, a lower stem yield. In order to assess which technique is most promising economically, the effect of harvest date on potential PAR interception and yield should be quantified. To do this, the crop growth model and the average weather data described above were used. It was assumed that a nonflowering cultivar sown on April 15 intercepted 99 percent of incident PAR from full canopy establishment until harvest.

In the current French and Hungarian hemp cultivars, stem growth ceases in the first half of September. According to the crop growth model, a hemp crop sown on April 15 would have intercepted 927 MJ m^{-2} PAR by September 15 (see Table 5.3). Advancing harvest date by thirty days from September 15 to August 16, in order to make field drying possible, would reduce intercepted PAR by 195 MJ m^{-2}. Delaying harvest date by thirty days to October 15 would increase intercepted PAR by a smaller amount: 123 MJ m^{-2}.

Radiation-Use Efficiency

Radiation-use efficiency (RUE) is defined here as the amount of dry matter produced per unit of intercepted PAR by a nonstressed crop (Monteith, 1977). The RUE of nonstressed crops depends on crop gross photosynthesis, maintenance respiration, and growth respiration (Charles-Edwards, 1982). Losses of dry matter during the growing season may cause an apparent reduction in RUE.

Table 5.3. Simulated effect of harvest date on accumulated light interception by the canopy of a hemp crop grown at a plant density of 64 m^{-2} and sown on April 15.

Harvest date	Sowing-harvest (days)	PAR intercepted by harvest (MJ m^{-2})
August 1	108	617
August 16	123	732
August 31	138	841
September 15	153	927
September 30	168	1000
October 15	183	1050

Analysis of the experiments conducted in the 1980s yielded RUEs of 2.0 to 2.2 g MJ^{-1} before flowering, and of 1.1 to 1.2 g MJ^{-1} after flowering (Meijer, van der Werf, Mathijssen et al., 1995). These RUE values are at the lower end of the range of values found for other C_3 crops; several factors are probably responsible. The crop gross photosynthesis of hemp is negatively affected during most of the growing season by the high extinction coefficient of the hemp canopy. Furthermore, growth respiration is probably relatively large in hemp, because lignin is being synthesized in the stem. After flowering, growth respiration increases further, because fat and protein are synthesized in the seed. Finally, losses of dry matter during the growing season are large as dead leaves are shed rapidly and many plants may die during the growing season as a result of self-thinning. In the experiments conducted in the 1980s, almost half of the plants had died before harvest in September, even at the lowest densities. Taken together, shed leaves and dead plants may represent up to 3 t ha^{-1} of dry matter, which is subject to biotic and abiotic degradation and is difficult to collect (Meijer, van der Werf, Mathijssen et al., 1995). Obviously these losses reduce the apparent RUE of hemp. We will focus here on two main factors involved in the low RUE of fiber hemp: flowering and self-thinning.

Flowering

Earlier studies had revealed that RUE postflowering was low in hemp (Meijer, van der Werf, Mathijssen et al., 1995). In subsequent experiments, RUE remained high (2.3 g MJ^{-1}) throughout September when flowering was prevented, but postflowering RUE was low (0.6 g MJ^{-1}) when flowering was not prevented (van der Werf, Haasken, and Wijlhuizen, 1994). These results were further corroborated in another experiment, in which a very late cultivar maintained a high RUE (2.2 to 2.3 g MJ^{-1}) until it flowered in September, whereas the RUEs of the other cultivars, which had flowered in August, were lower (1.9 g MJ^{-1}) in late August and early September (van der Werf, Wijlhuizen, and de Schutter, 1995).

A minor part of the postflowering decline in the RUE of hemp can be accounted for by larger losses of shed leaves and increased growth respiration due to the synthesis of fat and protein in the seed (van der Werf, Haasken, and Wijlhuizen, 1994). However, the decline seems to be caused in the first place by an important reduction of crop gross photosynthesis, probably as a result of senescence of the leaves. Breeding late-flowering cultivars may therefore offer scope for the prevention of the low RUE postflowering in hemp.

Self-Thinning

In an experiment conducted in 1988, the RUE prior to flowering was 2.2 g MJ^{-1} in a crop with an initial plant density of eighty-six plants m^{-2} and 2.0 g MJ^{-1} in a crop with an initial plant density of 342 m^{-2} (Meijer, van der Werf, Mathijssen et al., 1995). During the growing season more plants died in the high-density crop than in the low-density crop; as a result, in August and September the dry weight of dead plants was greater in the high-density crop. As dry matter of dead plants is inevitably degraded, measurements underestimate dead dry matter by an unknown amount. Thus, total dry matter production will be underestimated more at a high plant density than at a low plant density, and this seems a major cause of the lower RUE at the high plant density.

To examine this hypothesis further, hemp was grown at four plant densities (10, 30, 90, and 270 m^{-2}) to investigate the course of biomass yield and plant mortality during two growing seasons. It was established that interplant competition resulted in density-induced mortality, i.e., self-thinning (van der Werf, Wijlhuizen, and de Schutter, 1995). In a self-thinning hemp crop, an increase in biomass yield is accompanied by a reduction in plant density. An increase in the number of plants dying from self-thinning at 270 plants m^{-2} was associated with an increased amount of dead plant dry matter and a decline of the RUE, confirming the hypothesis outlined above. Unexpectedly, at ten plants m^{-2}, the amount of dead plant dry matter was also large, not as a result of plant mortality, but because the plants shed relatively large amounts of branches and leaves. Here too, the RUE declined as dead plant dry matter increased. At the two intermediate plant densities, little or no self-thinning took place, and little dead material was present. Apparent postflowering RUE was 1.9 g MJ^{-1} at the intermediate densities, 1.3 g MJ^{-1} at 270 plants m^{-2}, and 1.1 g MJ^{-1} at ten plants m^{-2}.

Plant mortality resulting from self-thinning can be prevented by ensuring that the plant density at emergence does not exceed the maximum plant density possible at the expected yield (van der Werf, Wijlhuizen, and de Schutter, 1995). For an aboveground dry matter yield of 15 t ha^{-1}, this plant density would be about 120 m^{-2}; at 20 t ha^{-1} it would be about 50 m^{-2}.

Dry Matter Partitioning

Fiber hemp is grown for the production of stem dry matter; within the stem the bark is more valuable than the core (van der Werf et al., 1994). Therefore, both a high proportion of stem in the aboveground dry matter and a high proportion of bark in the stem dry matter are desirable. Both levels of dry matter partitioning will be examined below.

Stem in the Aboveground Dry Matter

Data from Meijer, van der Werf, Mathijssen et al. (1995), van der Werf, Haasken, and Wijlhuizen (1994), van der Werf, Wijlhuizen, and de Schutter (1995), and van der Werf et al. (1995) on the partitioning of the aboveground dry matter to the inflorescence, leaves, and stem have been summarized in Table 5.4. In the experiments conducted in 1987, 1988, and 1989, the proportion of stem material in the aboveground dry matter of the monoecious cv. Fédrina 74 at harvest in September varied between 78 percent and 84 percent. The proportion of stem was not greatly affected by flowering and seed filling, because the increase in dry weight of the inflorescence was about as large as the decline in leaf dry weight (Meijer, van der Werf, Mathijssen et al., 1995). In 1989, cv. Fédrina 74 was compared with the dioecious cultivars Kinai unisexualis and Kenevir, both of which flowered about two weeks later than Fédrina 74. In that year, the proportion of the stem in the aboveground dry matter was 82 percent in Fédrina 74, 87 percent in Kinai unisexualis, and 86 percent in Kenevir (see Table 5.4).

Further experiments were conducted to examine the effect of flowering on the proportion of stem (van der Werf, Haasken, and Wijlhuizen, 1994). In Fédrina 74, the prevention of flowering by twenty-four hour days reduced inflorescence dry matter from 1.1 to 0.2 t ha^{-1} and increased leaf dry matter from 1.1 to 1.5 t ha^{-1} and stem dry matter from 10.7 to 13.4 t ha^{-1} (see Table 5.4). As a result, Fédrina 74 contained 89 percent of stem when flowering had been prevented, compared with 83 percent when it had flowered. In the dioecious cv. Kompolti Hybrid TC, which flowers about twenty days later than Fédrina 74, the prevention of flowering reduced inflorescence dry matter from 0.5 to 0.1 t ha^{-1} and increased leaf dry matter from 0.9 to 1.4 t ha^{-1} and stem dry matter from 13.2 to 15.9 t ha^{-1}. As a result, the proportion of stem was 92 percent when flowering had been prevented and 91 percent when flowering had occurred. When both cultivars flowered normally, inflorescence dry weight was larger in Fédrina 74 than in Kompolti Hybrid TC; when flowering was prevented, cultivar differences were smaller. There are probably two reasons for the large inflorescence in Fédrina 74. First, Fédrina 74 flowers earlier than Kompolti Hybrid TC, so allocation of dry matter to the inflorescence starts earlier. Second, Fédrina 74 is monoecious, so all plants invest dry matter in floral clusters and seeds, whereas Kompolti Hybrid TC is dioecious, containing about 50 percent of male plants, which die after flowering and contain a much smaller fraction of the aboveground dry matter in the inflorescence.

In another experiment with Kompolti Hybrid TC, the proportion of stem in the aboveground dry matter was found to be affected by plant density (van der Werf, Wijlhuizen, and de Schutter, 1995). At harvest in the middle of September it increased from 78 percent at ten plants m^{-2} to 86 percent at ninety plants m^{-2}; at 270 plants m^{-2} it was 85 percent (see Table 5.4). This increase in the proportion of stem resulted mainly from the dry weight of the inflorescence declining with increasing plant density.

In the same experiment, Kompolti Hybrid TC was compared with Kompolti Hyper Elite, a high bast fiber selection from cv. Kompolti and with Kozuhara zairai, a late-flowering cv. At harvest all three cultivars had high dry matter yields (about 18 t ha^{-1}). The proportion of stem in the aboveground dry matter was 86 percent in Kompolti Hybrid TC, 87 percent in Kompolti Hyper Elite, and 84 percent in Kozuhara zairai (see Table 5.4). The dry weight of leaves was 2.6 t ha^{-1} in the late-flowering cv. and 1.8 t ha^{-1} in the two other cultivars, and this was the major cause of the smaller proportion of stem in the late cv.

The level of soil nitrogen (van der Werf, van Geel et al., 1995) barely affected the proportion of stem in the aboveground dry matter. In early September, it was 89 percent at 80 kg ha^{-1} N and 88 percent at 200 kg ha^{-1} N.

Table 5.4. Dry matter in the inflorescence, in the leaves attached to the plant and in the stem, and the proportion of stem in the total dry matter, of living hemp plants harvested in September.

Year	Treatment	Dry matter (t ha-1)			
		Inflorescence	Leaves	Stem	Stem in total (%)
1987	Fédrina 74, 104 plants m^{-2}	0.25	1.23	7.58	84
1988	Fédrina 74, 86 plants m^{-2}	1.57	1.60	11.32	78
1989	Fédrina 74, 114 plants m^{-2}	0.68	2.26	13.35	82
	Kinai unisexualis	0.07	1.94	13.19	87
	Kenevir	0.06	2.05	12.69	86
1990-1991	Fédrina 74, 24-hour daylength	0.16	1.49	13.42	89
	Fédrina 74, normal daylength	1.13	1.12	10.65	83
	Kompolti Hybrid TC, 24-hour daylength	0.06	1.41	15.92	92
	Kompolti Hybrid TC, normal daylength	0.45	0.89	13.22	91
1991-1992	Kompolti Hybrid TC, 10 plants m^{-2}	1.16	1.84	10.80	78
	Kompolti Hybrid TC, 30 plants m^{-2}	0.84	2.04	14.50	83
	Kompolti Hybrid TC, 90 plants m^{-2}	0.58	1.85	15.10	86
	Kompolti Hybrid TC, 270 plants m^{-2}	0.45	1.76	12.90	85
	Kompolti Hyper Elite, 90 plants m^{-2}	0.50	1.75	15.20	87
	Kozuhara zairai, 90 plants m^{-2}	0.41	2.61	15.40	84
1991-1992	Kompolti Hybrid TC, 80 kg ha^{-1} N	0.26	0.99	10.35	89
	Kompolti Hybrid TC, 200 kg ha^{-1} N	0.32	1.25	11.30	88

Data summarized from Meijer et al. (1995); van der Werf, Haasken, and Wijlhuizen (1994); van der Werf, Brouwer et al. (1995); and van der Werf, van Geel, and Wijlhuizen (1995).

In conclusion, therefore, flowering date, plant density, and the proportion of male plants are the main factors affecting the proportion of stem in the aboveground dry matter of a hemp crop. The later a cultivar flowers, the smaller the fraction of the inflorescence and the larger the fraction of the leaves and the stem in the aboveground dry matter will be. The resulting effect of flowering date on the proportion of stem is variable so that, relative to an early cultivar, a late cultivar may contain a similar, smaller or larger proportion of stem. To obtain a high proportion of stem in the aboveground dry matter the crop should be grown at a the highest possible density not causing self-thinning. The more male plants are present, the larger the stem proportion.

Bark in the Stem Dry Matter

In a 1990 experiment (van der Werf et al., 1994), the proportion of bark in the stem dry matter (bark content) in September was higher at ninety plants m^{-2} than at ten plants m^{-2}, and higher in cv. Kompolti Hybrid TC than in cv. Fédrina 74.

Further experiments (van der Werf, Haasken, and Wijlhuizen, 1994) confirmed the difference in bark content between Fédrina 74 (31 percent) and Kompolti Hybrid TC (35 percent). In both cultivars, the bark content decreased during August and September, the decrease being slight in Kompolti Hybrid TC and more pronounced in Fédrina 74. For both cultivars, the decrease in bark content was associated with an increase in stem dry weight. When flowering had been artificially prevented, the increase in stem dry weight and the decrease in bark content were larger than when flowering had taken place normally.

The effect of plant density on bark content (van der Werf, Wijlhuizen, and de Schutter, 1995) was similar to its effect on the proportion of stem such that, at harvest in the middle of September, bark content increased from 33 percent at ten plants m^{-2} to 36 percent at ninety plants m^{-2}; at 270 plants m^{-2} it was 35 percent. In the same experiment, Kompolti Hybrid TC was compared with Kompolti Hyper Elite, and with Kozuhara zairai (van der Werf, Wijlhuizen, and de Schutter, 1995). At harvest, all three cultivars had high stem yields (15 t ha^{-1} of dry matter) and bark contents were 21 percent (Kozuhara zairai), 36 percent (Kompolti Hybrid TC), and 40 percent (Kompolti Hyper Elite).

The level of soil nitrogen affected the bark content: in early September it was 36 percent at 80 kg ha^{-1} N and 34 percent at 200 kg ha^{-1} N (van der Werf et al., 1995). The effect of soil nitrogen probably resulted mainly from a difference in plant density which had arisen as a result of self-thinning. In early September, plant density was 129 m^{-2} at 80 kg ha^{-1} N and 92 m^{-2} at 200 kg ha^{-1} N.

It seems, therefore, that cultivar and plant density are the main factors affecting the proportion of bark in the stem. To maximize bark content a high fiber cultivar should be grown at the highest possible density not causing self-thinning.

POTENTIAL YIELD

The simple crop growth model referred to previously (see Figure 5.1) was used to estimate the potential stem yield of fiber hemp. The major factors affecting the amount of light intercepted by a fiber hemp crop are the dates of sowing and harvesting. For this simulation, April 15 was chosen as a reference sowing date and the effect of earlier and later sowing dates was examined. Likewise, September 15 was used as a reference harvest date and the effect of varying harvest date was also examined.

As mentioned previously, radiation-use efficiency ranged from 0.6 to 2.3 g MJ^{-1}. Before flowering, and when no or little self-thinning occurred, the RUE was 2.2 to 2.3 g MJ^{-1}. For the simulations it was assumed that a hypothetical nonflowering cultivar was being grown, at a plant density (64 m^{-2}) which does not cause significant self-thinning unless yields exceed 20 t ha^{-1} of dry matter. Given these conditions, a RUE of 2.2 g MJ^{-1} is a realistic assumption.

In our experiments, the proportion of stem in the aboveground dry matter varied from 78 to 92 percent, and therefore it was difficult to choose a reference value. In the case of a nonflowering cultivar, the aboveground dry matter would consist of leaf and stem only. Based on the results obtained with the very late cv. Kozuhara zairai (van der Werf, Wijlhuizen, and de Schutter, 1995), the proportion of stem in the aboveground dry matter was assumed to be 84 percent in the simulation.

Figure 5.1. Schematic representation of the LINTUL crop growth model (Spitters, 1990) and of crop management decisions affecting the input relations.

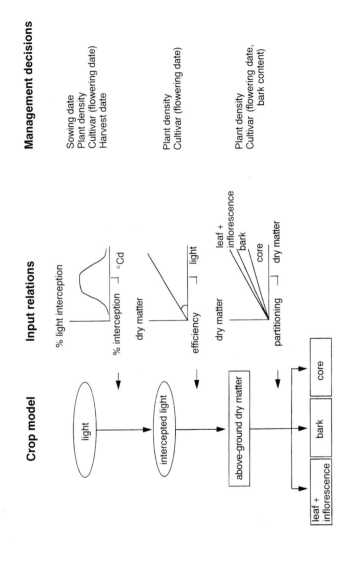

As pointed out earlier, bark content in the stem was affected by plant density, but more by cultivar. The plant density of 64 m^{-2} used in the model is close to the density at which bark content peaked. The highest bark content (40 percent), was found in the high bast fiber selection "Hyper Elite" from the cultivar Kompolti. Breeding research has shown that flowering date and fiber content are not necessarily linked (Hennink, 1994). Genetic variability for flowering date is large (de Meijer and Keizer, 1994) so that it should be feasible to breed a very late flowering, high bast fiber cultivar. A bark content of 40 percent was assumed for the hypothetical nonflowering cultivar.

According to the crop growth model, a nonstressed, nonflowering hemp cultivar sown on April 15 and harvested on September 15 would yield 17.1 t ha^{-1} of stem dry matter (see Figure 5.2). Sowing the crop on March 31 instead of April 15 would increase stem yield by 1.4 t ha^{-1}; sowing on March 16 would increase stem yield by 2.3 t ha^{-1} (see Figure 5.2). These yield increases are substantial, but should be weighed against the increased risk of frost damage. Sowing on April 30 instead of April 15 would reduce stem yield by 1.4 t ha^{-1}, sowing on May 15 would reduce stem yield by 3.3 t ha^{-1}. The more the sowing date is delayed, the more rapidly the potential stem yield drops, because light interception in the period of maximum incident radiation (May and June, see Table 5.1) is increasingly incomplete.

The yield increase obtained by delaying harvest date by fifteen or thirty days is almost identical to the yield increase obtained by advancing sowing date by fifteen or thirty days. Advancing the harvest date by fifteen days reduces the yield by 1.6 t ha^{-1}; advancing the harvest by thirty days reduces the stem yield by 3.6 t ha^{-1}. The effect of advancing harvest date on stem yield is slightly larger than the effect of delaying sowing date. The effect of simultaneous changes in sowing date and harvest date can be calculated from Figure 5.2 by summing the effects of both changes.

In conclusion, therefore, the dates of sowing and harvest both have large effects on the potential stem yield of a nonflowering hemp cultivar. Sowing earlier than April 15 can increase yield, certainly on soils that are not frost-prone. Harvesting in August instead of September decreases potential yield but will allow field drying. Delaying the harvest date offers scope for increased stem yields, but requires the breeding of very late-flowering cultivars.

Figure 5.2. The effect of advancing (−15, −30 days) or delaying (15, 30 days) sowing date or harvest date on the simulated stem yield of a hypothetical non-stressed, nonflowering hemp crop grown at a plant density of 64 m^{-2}. Reference date (0 days) for sowing: April 15, for harvest: September 15.

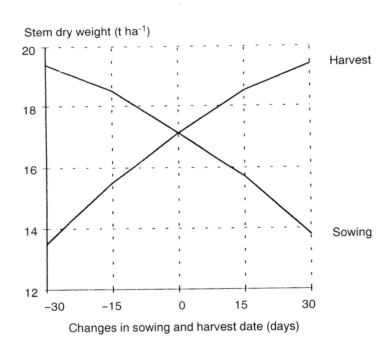

HEMP VERSUS KENAF

Interest in "new" fiber crops is increasing, for example, to replace cotton by a less polluting alternative or to relieve the pressure of the paper industry on remaining natural forests (Postel and Ryan, 1991). For the warmer parts of the world, ramie (*Boehmeria nivea* L.) may be an alternative to cotton, and kenaf seems to have excellent potential as an alternative source of paper pulp (Carberry et al., 1992). For the temperate regions of the globe, the perennial C_4-grass *Miscanthus sinensis* has been proposed as a raw material for paper (van der Werf et al., 1993; Wegener, 1993) and flax remains a valuable crop, providing raw material for textile but also for specialty paper. Although *M. sinensis* is highly productive and virtually disease-free, the length of its production cycle (ten years) makes it less attractive to arable farmers. Furthermore, the establishment of the crop is expensive and often hampered during winter by frosts and diseases. Flax, which like hemp is an annual bast fiber crop, continues to be grown in temperate climate zones, demonstrating that a market for bast fibers does exist. One of the major problems limiting the market potential of flax fiber is its high price (Judt, 1993). Under similar growing conditions, hemp yields are generally 50 to 100 percent higher than flax yields (Jordan, Lang, and Enfield, 1946; Meijer, van der Werf, van Roekel et al., 1995). As production costs are similar, hemp fiber should be cheaper than flax fiber and therefore have a better market potential. At these higher yield levels, hemp fiber will be coarser than flax fiber (van der Werf, Wijlhuizen, and de Schutter, 1995), which makes it less suitable for textile production but does not affect its quality as a raw material for paper.

The potential of hemp to produce raw material for paper in the temperate climate zone may be evaluated by comparing its crop physiological characteristics with those of kenaf, which has been proposed as a paper raw material for the subtropics and tropics.

The better a spring-sown crop grows at low temperatures (i.e., the lower its base temperature), the more rapidly it will establish its canopy. Hemp's base temperature for emergence and leaf appearance is $1°C$; for canopy establishment it is $2.5°C$ (van der Werf, Brouwer et al., 1995). For kenaf, the base temperature is $9°C$ for emergence (Angus et al., 1981), and $10°C$ for early growth (Carberry and Abrecht, 1990). For a temperate climate with a long period of low spring temperatures, hemp can be sown early and canopy establishment will be rapid. As a result, total light interception over the growing season will be large.

Before flowering, and when no self-thinning occurred, the RUE of hemp in our experiments was 2.2 to 2.3 g MJ^{-1}. This is slightly less than the RUE value of 2.4 g MJ^{-1} that Carberry and Muchow (1992) reported for kenaf, so that, in this respect, the two crops do not seem to differ much. This is not surprising, as the chemical composition of their stem is largely similar; both crops have a relatively large extinction coefficient and both crops lose dry matter during the growing season as dead leaves are shed rapidly.

Dry matter partitioning in hemp and kenaf is quite similar. According to Carberry and Muchow (1992) the proportion of the stem in the aboveground dry matter of kenaf varies from 83 to 89 percent, which is fully within the range we found for hemp. For kenaf, bark contents of up to 40 percent have been reported (Muchow et al., 1990), which again is similar to our results for hemp.

Radiation-use efficiency and dry matter partitioning are similar in hemp and kenaf. Due to its low base temperature, hemp is adapted to the cool springs of a temperate climate. Therefore, hemp seems an excellent candidate to fill the niche for an annual fiber crop in a temperate climate.

REFERENCES

Abel, E L. (1980). *Marihuana, the first 12,000 years*. New York: Plenum Press.
Angus, J. F., Cunningham, R. B., Moncur, M. W., and Mackenzie, D. H. (1981). "Phasic development in field crops. I. Thermal response in the seedling phase." *Field Crops Research*, 3:365-378.
Bakker, H. and van Kemenade, M. J. J. M. (1993). *Papier uit hennep van Nederlandse grond. Eindrapportage van vier jaar henneponderzoek: Samenvatting, conclusies en aanbevelingen.* [Paper from hemp grown in the Netherlands. Final report of four years of research on hemp: Summary, conclusions, and recommendations.] Wageningen, Netherlands: ATO-DLO.
Beaudelaire, C. (1860). "Le poëme du haschisch." In *Les Paradis artificiels* (1992 edition). Paris: Gallimard.
Bócsa, I. (1971). "Különleges célok elérése a kendernemesítésben." [The realization of specific breeding goals in hemp breeding.] *Rostnövények*, 29-34.
Bócsa, I. (1995). "Die Hanfzüchtung in Ungarn: Zielersetzungen, Methoden und Ergebnisse." [The breeding of hemp in Hungary: Objectives, methods, and results.] In *Bioresource Hemp Symposium Reader*, Second Edition, pp. 200-215. Frankfurt, Germany: Nova-Institut.
Carberry, P. S. and Abrecht, D. G., (1990). "Germination and elongation of the hypocotyl and radicle of kenaf (*Hibiscus cannabinus* L.)." *Field Crops Research*, 24:227-240.
Carberry, P. S. and Muchow, R. C., (1992). "A simulation model of kenaf for assisting fiber industry planning in northern Australia. III. Model description and validation." *Australian Journal of Agricultural Research*, 43:1527-1545.

Carberry, P. S., Muchow, R. C., Williams, R., Sturz, J. D., and McCown, R. L., (1992). "A simulation model of kenaf for assisting fiber industry planning in northern Australia. I. General introduction and phenological model." *Australian Journal of Agricultural Research*, 43:1501-1513.

Charles-Edwards, D. A. (1982). *Physiological determinants of crop growth.* Sydney, Australia: Academic Press.

Deferne, J. L. and Pate, D. W. (1996). "Hemp seed oil: A source of valuable fatty acids." *Journal of the International Hemp Association*, 3:1-7.

de Groot, B., van Dam, J. E. G., van der Zwam, R. P., and van 't Riet, K. (1994). "Simplified kinetic modelling of alkaline delignification of hemp woody core." *Holzforschung*, 48:207-214.

de Groot, B., van Zuilichem, D. J., and van der Zwam, R. P. (1988). "The use of non-wood fibers for pulping and papermaking in the Netherlands." In Vol. 1, *Proceedings of the 1988 International Non-Wood Fiber Pulping and Papermaking Conference.* Beijing, China, pp. 216-222.

de Jonge, L. J. A. (1944). *Hennepteelt in Nederland.* [The cultivation of hemp in the Netherlands.] Wageningen, Netherlands: Veenman en zonen.

de Maeyer, E. A. A. and Huisman, W. (1994). "New technology to harvest and store fiber hemp for paper pulp." *Journal of the International Hemp Association*, 1:38-41.

Dempsey, J. M. (1975). "Hemp." In *Fiber crops.* Gainesville, FL: University of Florida Press, pp. 46-89.

du Bois, W. F. (1982). "Hennep als grondstof voor de papierindustrie." [Hemp as a raw material for the paper industry.] *Bedrijfsontwikkeling*, 13:851-856.

du Bois, W. F. (1984). *Hennep als grondstof voor papier.* [Hemp as a raw material for paper.] Wageningen, Netherlands: Institute for the Preservation and Transformation of Agricultural Products IBVL, report number 482.

FAO. (1995). FAO Yearbook, Production 1994, Vol. 48, FAO Statistics series. Rome, Italy: Food and Agriculture Organization of the United Nations.

Fournier, G., Richez-Dumanois, C., Duvezin, J., Mathieu, J-P., and Paris, M. (1987). "Identification of a new chemotype in *Cannabis sativa*: Cannabigerol-dominant plants, biogenetic and agronomic prospects." *Planta Medica*, 53:277-280.

Goloborod'ko, P. (1995). "Hemp research and growing in Ukraine." *Journal of the International Hemp Association*, 2:35-37.

Grenikov, A. S. and Tollochko, T. (1953). *Cultivation of hemp* [in Russian]. Moscow, USSR: State Editors of Agricultural Literature.

Hanson, J. (1980). "An outline for a U.K. hemp strategy." *The Ecologist*, 10:260-263.

Hennink, S. (1994). "Optimisation of breeding for agronomic traits in fiber hemp (*Cannabis sativa* L.) by study of parent-offspring relationships." *Euphytica*, 78:69-76.

Herer, J. (1985). *Hemp and the marijuana conspiracy: The emperor wears no clothes.* Van Nuys, CA: HEMP Publishing.

Heuser, O. (1927). *Hanf und Hartfaser.* Berlin, Germany: Julius Springer-Verlag.
Jordan, H. V., Lang, A. L., and Enfield, G. H. (1946). "Effects of fertilizers on yield and breaking strengths of American hemp, *Cannabis sativa.*" *Journal of the American Society of Agronomy,* 38:551-563.
Judt, M. (1993). "Non-wood plant fibers: Will there be a comeback in paper-making?" *Industrial Crops and Products,* 2:51-57.
Kok, C. J., Coenen, G. C. M., and de Heij, A. (1994). "The effect of fiber hemp (*Cannabis sativa* L.) on selected soil-borne pathogens." *Journal of the International Hemp Association,* 1:6-9.
Mathieu, J. P. (1980). "Chanvre." *Techniques Agricoles,* 5:1-10.
McDougall, G. J., Morrison, I. M., Stewart, D., Weyers, J. D. B., and Hillman, J. R. (1993). "Plant fibers: Botany, chemistry and processing for industrial use." *Journal of the Science of Food and Agriculture,* 62:1-20.
Meijer, W. J. M., van der Werf, H. M. G., Mathijssen, E. W. J. M., and van den Brink, P. W. M. (1995). "Constraints to dry matter production in fiber hemp (*Cannabis sativa* L.)." *European Journal of Agronomy,* 4:109-117.
Meijer, W. J. M., van der Werf, H. M. G., van Roekel, G. J., de Meijer, E. P. M., and Huisman, W. (1995). "Fiber hemp: Potentials and constraints." In *Opportunities and profits: Proceedings Special Crops Conference,* Calgary, Canada, July 1995. Alberta Agriculture, Food and Rural Development, pp. 67-79.
Meijer de, E. P. M. (1994). "Diversity in *Cannabis.*" PhD thesis, Wageningen Agricultural University.
Meijer de, E. P. M. (1995). "Fiber hemp cultivars: A survey of origin, ancestry, availability and brief agronomic characteristics." *Journal of the International Hemp Association,* 2:66-73.
Meijer de, E. P. M. and Keizer, L. C. P. (1994). "Variation of *Cannabis* for phenological development and stem elongation in relation to stem production." *Field Crops Research,* 38:37-46.
Milford, G. F. J., Pocock, T. O., and Riley, J. (1985). "An analysis of leaf growth in sugar beet. I. Leaf appearance and expansion in relation to temperature under controlled conditions." *Annals of Applied Biology,* 106:163-172.
Milford, G. F. J., Pocock, T. O., Riley, J., and Messem, A. B., (1985). "An analysis of leaf growth in sugar beet. III. Leaf expansion in field crops." *Annals of Applied Biology,* 106:187-203.
Monteith, J. L. (1977). "Climate and the efficiency of crop production in Britain." *Philosophical Transactions of the Royal Society of London, B,* 281:277-294.
Muchow, R. C., Sturtz, J. D., Spillman, M. F., Routley, G. E., Kaplan, S., Martin, C.C., and Bateman, R. J. (1990). "Agronomic studies on the productivity of kenaf (*Hibiscus cannabinus* L. cv. Guatemala 4) under rainfed and irrigated conditions in the Northern Territory." *Australian Journal of Experimental Agriculture,* 30:395-403.
Pesticides Trust. (1990). *King Cotton and the pest.* London, United Kingdom: The Pesticides Trust.

Pimentel, D., McLaughlin, L., Zepp, A., Lakitan, B., Kraus, T., Kleinman, P., Vancini, F., Roach, W. J., Graap, E., Keeton, W. S., and Selig, G. (1991). "Environmental and economic impacts of reducing U.S. agricultural pesticide use." In D. Pimentel (Ed.), *CRC handbook of pest management in agriculture*, Second Edition, pp. 679-718. Boca Raton, FL: CRC Press.

Pleydell-Bouverie, J. (1994). "Cotton without chemicals." *New Scientist*, September 24, 1994:25-29.

Postel, S., Ryan, J. C. (1991). "Reforming forestry." In Linda Starke (Ed.), *State of the world 1991*, pp. 75-92. New York, London: W. W. Norton & Company.

Pounds, N. J. G. (1979). *An historical geography of Europe 1500-1840*. Cambridge, United Kingdom: Cambridge University Press.

Rabelais, F. (1546). "Le tiers livre des faicts et dicts heroiques du bon Pantagruel." In *Rabelais Oeuvres complètes* (1955 edition), pp. 316-514. Paris, France: Gallimard.

Schultes, R. E. (1970). "Random thoughts and queries on the botany of *Cannabis*." In C. R. B. Joyce and S. H. Curry (Eds.), *The botany and chemistry of Cannabis*, pp. 11-38. London, United Kingdom: J & A Churchill.

Senchenko, G. I. and Demkin, A. P. (1972). "Methods for increasing yield and quality of hemp [in Russian]. *Journal of Farm Mechanization Science*, Moscow, 9:52-59.

Senchenko, G. I. and Timonin, M. A. (1978). *Hemp* [in Russian]. Moscow, USSR: Kolos.

Smit, A. L. (1989). *Overzaaien van suikerbieten*. [Resowing of sugar beet.] PAGV Report no. 91, Lelystad, Netherlands.

Smit, A. L. (1993). "The influence of sowing date and plant density on the decision to resow sugar beet." *Field Crops Research*, 34:159-173.

Spitters, C.J.T. (1990). "Crop growth models, their usefulness and limitations." *Acta Horticulturae*, 267:349-368.

van der Werf, H. M. G. (1994). "Crop physiology of fiber hemp *(Cannabis sativa L.)*." PhD thesis, Wageningen Agricultural University, Wageningen, Netherlands.

van der Werf, H. M. G., Brouwer, K., Wijlhuizen, M., and Withagen, J. C. M. (1995). "The effect of temperature on leaf appearance and canopy establishment in fiber hemp *(Cannabis sativa L.).*" *Annals of Applied Biology*, 126:551-561.

van der Werf, H. M. G., Haasken, H. J., and Wijlhuizen, M. (1994). "The effect of daylength on yield and quality of fiber hemp *(Cannabis sativa L.).*" *European Journal of Agronomy*, 3:117-123.

van der Werf, H. M. G., Harsveld van der Veen, J. E., Bouma, A. T. M., and Cate, M. ten. (1994). "Quality of hemp *(Cannabis sativa L.)* stems as a raw material for paper." *Industrial Crops and Products*, 2:219-227.

van der Werf, H. M. G., Meijer, W. J. M., Mathijssen, E. W. J. M., and Darwinkel, A. (1993). "Potential dry matter production of *Miscanthus sinensis* in the Netherlands." *Industrial Crops and Products*, 1:203-210.

van der Werf, H. M. G. and van den Berg, W. (1995). "Nitrogen fertilization and sex expression affect size variability of fiber hemp (*Cannabis sativa* L.)." *Oecologia*, 103:462-470.

van der Werf, H. M. G., van Geel, W. C. A., van Gils, L. J. C., and Haverkort, A. J., (1995). "Nitrogen fertilization and row width affect self-thinning and yield of fiber hemp (*Cannabis sativa* L.)." *Field Crops Research*, 42: 27-37.

van der Werf, H. M. G., van Geel, W. C. A., and Wijlhuizen, M. (1995). "Agronomic research on hemp (*Cannabis sativa* L.) in the Netherlands, 1987-1993." *Journal of the International Hemp Association*, 2:14-17.

van der Werf, H. M. G., Wijlhuizen, M., and de Schutter, J. A. A. (1995). "Plant density and self-thinning affect yield and quality of fiber hemp (*Cannabis sativa* L.)." *Field Crops Research*, 40:153-164.

van Roekel, G.J. (1994). "Hemp pulp and paper production." *Journal of the International Hemp Association*, 1:12-14.

van Roekel, G. J., Lips, S. J. J., op den Kamp, R. G. M., and Baron, G. (1995). Extrusion pulping of true hemp bast fiber (*Cannabis sativa* L.)." *Proceedings of TAPPI pulping conference, October 6-10, 1995*, Chicago, pp. 477-485.

Vavilov, N. I. (1926). "Centers of origin of cultivated plants." In V. F. Dorofeyev (Ed.), *Origin and geography of cultivated plants*, pp. 22-135. Cambridge, United Kingdom: Cambridge University Press.

Wegener, G. (1993). "Pulping innovations in Germany." *Industrial Crops and Products*, 1:113-117.

Chapter 6

A Survey of Hemp Diseases and Pests

John M. McPartland

INTRODUCTION

This survey reviews common diseases and pests of the fiber hemp plant, *Cannabis sativa* L. This species is also a source of marijuana and hashish (along with *Cannabis indica* and *Cannabis afghanica*), but drug crop plants are not the subject of this survey. Drug crops share many problems with fiber crops, but the two crops also have different problems. This difference is partially due to genetic variations. The fiber biotype produces large, long stalks, which become very attractive to stalk-boring insects and stalk-canker fungi. The drug biotype produces large flowers, which become very attractive to budworms and gray mold. Rothschild, Rowan, and Fairbairn (1977) found moths easily distinguished between plants with high- and low-THC levels. McPartland (1992) reported different disease prevalences between fiber and drug *Cannabis*. Noviello et al. (1990) found that fiber cultivars are more resistant to the fungus *Fusarium oxysporum* f. sp. *cannabis* than drug plants. Part of this difference in resistance was genetic.

Different problems also arise due to horticultural differences. Fiber crops are grown in tight proximity. Canopy closure occurs early in high-density stands, increasing humidity around stalks and

---

Management of Hemp Diseases and Pests, with Emphasis on Biocontrol, will be published in 1999. If you desire a book announcement when that work is published, please send your name and address to Dr. John McPartland, 53 Washington Street, Middlebury, Vermont 05753.

predisposing to fungal diseases. Canopy closure embraces insects in a safe, protected environment. In contrast, drug crops and seed crops grow in wide-open rows where the canopy may never close. These crops are more attractive to sunlight-loving pests, such as flea beetles and birds. Drug crops are often grown indoors (glasshouses or growrooms), and these situations become susceptible to pests that reproduce rapidly, such as spider mites and aphids. Spider mites and aphids are less of a problem outdoors, because outdoors these pests are destroyed by many natural predators and parasites. Predators and parasites are rarely found indoors unless they are introduced by the grower.

Many current authors claim hemp is problem-free (Herer, 1991; Conrad, 1994; Rosenthal, 1994). None of these authors has ever cultivated a fiber crop. In reality, hemp is not pest-*free*, it is pest-*tolerant*; many problems arise in *Cannabis*, but these problems rarely cause catastrophic damage. However, diseases and pests cause small losses that may accumulate over time to significant numbers. Agrios (1988) estimates that 13 percent of fiber crops are lost to insects, 11 percent are lost to diseases, and 7 percent are lost to weeds and other organisms. In addition to these losses in the field, Pimentel et al. (1991) adds another 9 percent in *postharvest* storage losses. Add these numbers and you reach 40 percent.

This review is sectioned taxonomically, beginning with the most common pests—insects. Pests and diseases in each section are presented in their approximate order of economic impact. Common names used here are standards established by Stoetzel (1989) and McPartland (1991).

INSECT PESTS OF STALK AND ROOTS

Nearly 300 insects have been described on *Cannabis* but very few cause appreciable crop losses (McPartland, 1996b). In hemp crops the most serious insect pests are stem borers, and the worst stem borers are caterpillars. The most economically damaging caterpillars are European corn borers *(Ostrinia nubilalis)*, and hemp borers *(Grapholita delineana)*.

European corn borers (ECBs) are native to eastern Europe. Before the introduction of maize *(Zea mays* L.) into Europe, the original host plants of ECBs were *Cannabis* and *Humulus* (hops). Thereafter, ECBs switched to maize (Nagy, 1986). But ECBs continue to attack *Cannabis*. They cause stalk cankers that are swollen and structurally weak. Stalks supporting flowering tops often break at cankers. Larvae boring into smaller branches cause wilting of distal plant parts. Under heavy infestations entire plants collapse. Emchuk (1937) states five to twelve larvae can destroy a hemp plant. The "entry holes" where ECBs bore into stalks become open wounds, providing access for fungi and other insects. One to four generations of ECBs may arise per growing season, depending on the latitude and local weather. ECBs that hatch late in the growing season may not bore into stalks. Instead, they infest flowering tops, where they eat leaves and flowers, spin webs, and scatter feces. Summers with high humidity and little wind favor ECB egg-laying, egg survival, and larval survival.

Hemp borers (HBs) *Grapholita delineana* are smaller than ECBs. HBs cause stalk cankers that are similar to ECB cankers, but smaller. HBs tend to infest the upper third of the plant (Nagy, 1967), whereas 91 percent of ECB galls are located in the lower three-quarters of hemp stalks (Nagy, 1959). Late-season HBs are very destructive in flowering tops; they are also called hemp leaf rollers and hemp seedeaters. HBs have been known to destroy 80 percent of a crop's flowering tops (Kryachko et al., 1965), and 41 percent of seed crops (Bes, 1978). Each HB larva consumes an average of sixteen *Cannabis* seeds (Smith and Haney, 1973). Baloch, Mushtaque, and Ghani (1974) determined that forty larvae will kill a *Cannabis* seedling (15 to 25 cm tall) in ten days. As little as ten larvae per plant cripple growth and seed production. One to three generations of HBs may arise per growing season, depending on the latitude.

Other caterpillars that bore into hemp stems include the goat moth *(Cossus cossus)*, common stalk borer *(Papaipema nebris)*, burdock borer *(Papaipema cataphracta)*, and a Japanese ghost moth *(Endocylyta excrescens)*. Some of these species also bore into roots.

Many beetle larvae also bore stalks and roots. The worst of these is the hemp flea beetle *(Psylliodes attenuata)*. Flea beetle grubs bore into roots, and adults eat leaves and flowering tops. They are a serious problem in eastern Europe and China. Usually two generations arise per year (Angelova, 1968). Other flea beetle pests include *Psylliodes punctulata, Phyllotreta nemorum, Phyllotreta atra, Podagrica aerata, Podagrica malvae, Chaetocnema hortensis,* and *Chaetocnema concinna.* Many of these flea beetles are easily confused with the hemp flea beetle.

Several weevils and curculios cause problems in hemp. Like flea beetles, the adults chew small holes in leaves and grubs feed on pith within stems or roots. The worst is the cabbage curculio, sometimes called the hemp curculio *(Ceutorhynchus rapae),* a European pest now found in North America. Tremblay (1968) reports three other *Ceutorhynchus* species infesting Italian hemp—*C. pleurostigma, C. quadridens,* and *C. roberti.* Goidànich (1928) cites seven other Curculiónidae on Italian hemp—*Ceutorhynchus suleicollis, Gymnetron labile, Gymnetron pascuorum, Polydrosus sericeus, Sitona humeralis, S. lineatus,* and *S. sulcifrons.* The hemp weevil *(Rhinocus pericarpius),* was considered the most injurious pest of hemp in Japan (Harada, 1930).

Grubs of at least two species of tumbling flower beetles (Mordéllidae) feed on stem pith—*Mordellistena micans (= Mordellistena cannabisi)* and *Mordellistena parvula. M. parvula* feeds in the upper parts of plants, within narrow branches, petioles, and even central leaf veins, whereas *M. micans* feeds in the lower end of stalks.

Hemp longhorn beetles *(Thyestes gebleri)* are large, black-and-white striped beetles with striped antennae nearly as long as their bodies. Grubs feed within larger stems and stalks, ejecting excrement at intervals through frass holes. They occur in east Asia; one generation arises per year.

The white root grubs of several kinds of scarab beetles gnaw at roots. They include European chafers *(Melolontha hippocastani, Melolontha melolontha,* and *Melolontha vulgaris),* the Japanese beetle *(Popillia japonica),* and *Maladera holosericea.*

Ants and white ants (termites) often curse *Cannabis* in semitropical climates. Cherian (1932) cites the fire ant *Solenopsis geminata* tunneling into roots of mature plants. Cherian also cites the termite *Odontotermes obesus* as a major pest of Indian marijuana. Clarke (personal communication, 1996) found termites in Mexico and equatorial Africa that hollowed stems up to the level of flowers.

INSECT PESTS OF LEAVES, FLOWERS, AND SEEDS

The literature reports dozens of leaf-eating caterpillars on *Cannabis*. This list includes many improbable pests (e.g., *Malacosoma americanum*, *Amyna octo*, and *Acherontia atropos*). Some improbable pests are probably incidental migrants—such as the American dagger *(Acronicta americana)* that I found on wild hemp growing under trees *(A. americana* is arboreal and probably fell from overhead). Several species of leaf-eating caterpillars are cited often enough to be considered serious pests: the silver Y-moth *(Autographa gamma)*, dot moth *(Melanchra persicariae)* cabbage moth *(Mamestra brassicae)*, garden tiger moth *(Arctia caja)*, and beet webworm *(Loxostege sticticalis)*. Some caterpillars accumulate in large numbers and crawl *en masse* across fields, defoliating everything in their path. They are called armyworms; species cited on hemp include the beet armyworm *(Spodoptera exigua)*, and bertha armyworm *(Mamestra configurata)*.

Caterpillars most damaging to flowers and seeds are noctùid budworms, such as the cotton bellworm, *Heliothis armigera* (sometimes called by its original name, *Helicoverpa armigera*). *H. armigera* is very prolific, producing one to six generations per year depending on latitude. Female moths lay over 1,000 eggs. *H. armigera* commonly infests cotton, maize, tobacco, and chickpeas. Cherian (1932) reports that 100 larvae of *H. armigera* could eat a pound of *Cannabis* per day! Two other budworms cited as pests on hemp are *Helicoverpa zea* and *Heliothis viriplaca*.

Hemp flea beetles *(Psylliodes attenuata)*, already mentioned for their root-boring larvae, actually cause much more damage as leaf-eating adults (Ragazzi, 1954). Since flea beetles are small and leap frequently, they do not eat much in one place. Leaf damage consists of many small, round to irregular holes, between veins. Leaves of heavily infested plants can become completely skeletonized. The adults of other beetles and weevils may also become serious pests, such as the aforementioned cabbage curculio, *Ceutorhynchus rapae* (Goidànich, 1928; Ferri, 1959a/1961c).

Leafminers can be a problem. Like miniature coal miners, leafminers tunnel their way through tissue within leaves. Most leafminers are maggots—members of the fly family Agromyzidae. Each leafminer species makes tunnels with the same identifiable "signature"—either linear, serpentine, or blotch-like. The most common leafminers on hemp are *Liriomyza strigata, Phytomyza horticola, Liriomyza cannabis, Agromyra strigata, Phyllotreta nemorum, Agromyza reptans,* and *Liriomyza eupatorii* (Anonymous, 1940; Ciferri and Brizi, 1955; Ferri, 1959a).

Leaves and flowering tops are also infested by insects with piercing-sucking mouthparts, such as aphids, whiteflies, leafhoppers, mealybugs, scales, bugs, and thrips. Destructive aphids include the green peach aphid *(Myzus persicae)*, black bean aphid *(Aphis fabae)*, hemp louse or bhang aphid *(Phorodon cannabis)*, and the hops aphid *(Phorodon humuli)*. Except for *P. cannabis*, these insects are polyphagous and have complicated life cycles that may span more than one host plant species. *P. cannabis* only infests *Cannabis*. It was originally described in 1860 by Passerini. *P. cannabis* vectors the hemp streak virus (Goidànich, 1955), hemp mosaic virus, hemp leaf chlorosis virus (Ceapoiu, 1958), cucumber mosaic, hemp mottle virus, and alfalfa mosaic virus (Schmidt and Karl, 1970).

Leafhoppers and their relatives cause minor damage, but there are many of them: the glasshouse leafhopper *(Zygina pallidifrons)*, redbanded leafhopper *(Graphocephala coccinea)*, potato leafhopper *(Empoasca fabae)*, flavescent leafhopper *(Empoasca flavescens)*, spittlebug *(Pilaenus spumarius)*, and several Asian planthoppers—*Geisha distinctissima, Ricania japonica, Stenocranus qiandainus,* and *Eurybrachys tomentosa*. The greenhouse whitefly *(Trialeurodes vaporariorum)*, sweet potato or tobacco whitefly *(Bemisia tabaci)*, and silverleaf whitefly *(Bemisia argentifolii)* sometimes infest semitropical crops, but they primarily cause problems in glasshouse crops.

Mealybugs and scales suck plant sap, like their homopteran cousins in the suborder Sternorrhyncha (the aphids and whiteflies). All homopteran pests will gum-up plant surfaces with honeydew. Honeydew attracts ants and supports the growth of sooty mold. Prominent mealybug pests include the cottonycushion scale *(Icerya purchasi)*, longtailed mealybug *(Pseudococcus longispinus)*, European fruit scale *(Parthenolecanium corni)*, and white peach scale *(Pseudaulacaspis pentagona)*.

The worst insects are tarnished plant bugs *(Lygus lineolaris)*, green stink bugs *(Nezara viridula)*, false chinch bugs *(Nysius ericae)*, potato bugs *(Calocoris norrvegicuss)*, and a new pest in Amsterdam, *Liocoris tripustulatus*. Some of these species also feed on pollen and immature seeds (Goidànich, 1928; Ragazzi, 1954; Ferri, 1959a).

Thrips have rasping mouth parts which they use to suck sap. Thrips are becoming a problem in modern glasshouses that use rockwool and hydroponics. In old soil-floored glasshouses, watering with a hose kept floors damp. This encouraged the fungus *Entomophthora thripidum*, which lives in damp soils. *E. thripidum* infects thrips when they drop to the ground to pupate. Now, there is no damp soil, no fungus, no natural biocontrol. Bad actors include the onion thrips *(Thrips tabaci)*, greenhouse thrips *(Heliothrips haemorrhoidalis)*, and the host-specific marijuana thrips *(Oxythrips cannabensis)*. *T. tabaci* can transmit hemp streak virus (Ceapoiu, 1958) and the *Cannabis* pathogen Argentine sunflower virus (Traversi, 1949).

Grasshoppers and their orthopteran allies—locusts, crickets, and cockroaches—eat round, smooth-edged holes in leaves. Heavy swarms of locusts in west Africa can strip plants to stalks in a matter of minutes (Clarke, personal communication, 1994). Clarke also reports large grasshoppers in Mexico biting through stems of young seedlings to topple them over so the pests can easily feed on leaves. Cited pests include the sprinkled locust *(Chloealtis conspersa)*, *Dichroplus maculipennis*, *Atractomorpha crenulata*, *Hieroglyphus nigrorepletus*, *Chrotogonus saussurei*, and *Tettigonia cantans*.

Seedlings have their own special insect pests. Seedlings are cut down by cutworms *(Agrotis ipsilon, Spodoptera litura,* and *Spodoptera exigua)*, crickets *(Gryllus desertus* and *Gryllus chinensis)*, hemp flea beetles *(Psylliodes attenuata)*, and root maggots *(Delia platura* and, according to one researcher, *Delia radicum)*.

NONINSECT PESTS

The most prominent noninsect pests are mites. The most serious mites are spider mites—the two-spotted spider mite *(Tetranychus urticae)* and the carmine spider mite *(Tetranychus cinnabarinus)*. *T. urticae* prefers warm regions with temperatures around 30° C and low humidity, whereas *T. cinnabarinus* likes hot regions and hot glasshouses with temperatures above 35° C. Adults of both species are oval to pear-shaped, 0.4 to 0.5 mm long, with eight legs; *T. urticae* mites are mostly light green but turn orange-red as they diapause; *T. cinnabarinus* adults are mostly orange-red but in cooler climates turn green—so the two species can be hard to differentiate.

Damage by spider mites can be confused with aphid damage or sudden fungal wilts. Late-season hemp borers and assorted budworms hide in webbing that is mite-like.

The hemp russet mite *(Aculops cannabicola)* is becoming a serious pest. Unlike spider mites, eriophyid mites are pale beige, elongate, soft-bodied and wormlike, with only two pairs of legs *A. cannabicola* is difficult to eradicate, resistant to many pesticides, and suffers no known biocontrol predators or parasites.

Slugs can be vexing pests in damp corners of the globe. Species such as *Deroceras reticulatum, Limax maximus,* and *Arion* species destroy seedlings and damage young plants. Slugs prefer wet weather and crops planted adjacent to meadows, pastures, or woods. They usually feed at night.

Any ornithologist worth his or her weight in binoculars knows birds devour *Cannabis* seeds. Early reports from Kentucky describe the passenger pigeon *(Ectopistes migratorius)* feeding on hemp seeds (Allen, 1908). Sorauer (1958) cites many seedeating European birds including the hemp linnet *(Carduelis cannabina),* magpie *(Pica pica),* starling *(Sturnus vulgaris),* common purple grackle *(Quiscalus quiscula),* tree sparrow *(Passer montanus),* English sparrow *(P. domesticas),* nuthatch *(Sitta europaea),* lesser spotted woodpecker *(Dendrocopus minor),* and turtledove *(Streptopelia turtur).*

Mammals that eat roots, shoots, leaves, or seeds are found in the Order Rodentia (mice, moles, field voles, gophers, and groundhogs/woodchucks), Order Lagomorpha (cottontails), and Order Artiodactyla (deer). Some humans *(Homo sapiens)* can also be pretty destructive.

FUNGAL DISEASES

Most diseases of hemp are caused by fungi. Over 420 fungal taxa have been described in the scientific literature as *Cannabis* pathogens. Many of these names are taxonomic synonyms (McPartland, 1995e); other species cited in the literature are misidentifications (McPartland, 1995a). After a name-by-name review, McPartland (1992) determined the 420+ taxa appearing in *Cannabis* literature represent eighty-eight true species of *Cannabis* pathogens. Only a few cause economic crop losses.

By far the worst disease is gray mold, caused by *Botrytis cinerea*. The fungus has a complicated life cycle. It produces gray spores that grow in grape-like clumps—thus the Latin name *Botrytis cinerea*. These spores have also been called *Botrytis infestans* by Saccardo in 1887, *Botrytis felisiana* by Massalongo in 1899, and *Botrytis vulgaris* by Ferraris in 1935. The fungus occasionally produces other spores (ascospores) that look completely different. These spores have a different name—*Botryotinia fuckeliana*—but it is the same fungus. The gray mold fungus thrives in regions with high humidity (>60 percent RH) and cool to moderate temperatures (20° to 24° C). Under these conditions it can reach epidemic proportions and completely destroy a *Cannabis* crop in a week (Barloy and Pelhate, 1962). In fiber crops, gray mold often arises as a stalk rot. Enzymes released by the fungus turn stalks into soft shredded cankers. Stalks may snap at the cankers. Gray mold also arises as a bud rot of flowering tops. Large moisture-retaining female buds are most susceptible. The whole flower becomes covered in a gray fuzz, then turns into slime. Plants become most susceptible after flowers have peaked and pistils start to wilt.

Seedlings also succumb to the gray mold fungus, but disease in seedlings is called *damping off*. Seedlings attacked by damping off quickly wilt and fall over dead. Or seedlings die before they emerge from the soil (called preemergent damping off). Damping off is caused by many fungi: *Botrytis cinerea, Rhizoctonia solani, Macrophomina phaseolina*, and several *Fusarium* species *(F. solani, F. oxysporum, F. sulphurem, F. avenaceum,* and *F. graminearum)*. But the most common causes of damping off are two *Pythium* species (technically they are oömycetes, not fungi), *P. aphanidermatum* and *P. ultimum*.

Stalk cankers are common, and *Sclerotinia sclerotiorum* is a very common cause. The disease caused by *S. sclerotiorum* has many common names around the world—cottony soft rot, watery soft rot, stem rot, white mold, and gray rot. Europeans call it hemp canker, where some consider it the number-one scourge of hemp cultivation (Rataj, 1957) or number two behind gray mold (Termorshuizen, 1991). *S. sclerotiorum* sclerotia can be confused with sclerotia formed by the gray mold fungus (in old literature called *Sclerotinia fuckeliana*).

Many *Fusarium* species cause stalk cankers—*F. sulphureum, F. graminearum, F. lateritium, F. sambucinum, F. avenaceum,* and *F. culmorum* (Ferri, 1961b; McPartland and Cubeta, 1997). Most of these species produce *Gibberella* teleomorphs. *Fusarium* cankers often exhibit a red discoloration.

Many other fungi cause stalk cankers. The first disease ever identified on hemp was a canker fungus, *Sphaeria cannabis* (Schweinitz, 1832). Examination of Schweinitz's original specimen reveals it is a strain of the common fungus *Botryosphaeria obtusa* (McPartland, 1995b). Saccardo found *Hymenoscyphus (Helotium) herbarum* and *Leptosphaeria acuta* fruiting on *Cannabis* stems near Padova, Italy. Four species of *Cladosporium* reportedly cause *Cannabis* cankers, the most common being *C. herbarum* (Curzi and Banbaini, 1927). *Ophiobolus cannabinus* was discovered on hemp stalks near Parma by Passerini in 1888. *Coniothyrium cannabinum* has been collected in Italy (Curzi, 1927; Bestagno-Biga, Ciferri, and Bestagno, 1958). *Phomopsis cannabina,* named by Curzi in 1927, may be identical to *Phomopsis arctii* (Saccardo) Traverso, according to McPartland (1995e). *Rhizoctonia solani* and a binucleate *Rhizoctonia* species cause "sore shin," a disease of the lower stem and roots (McPartland and Cubeta, 1997).

Root rot is difficult to diagnose because the damage is underground where you cannot see it. The most common root rot is caused by *Fusarium solani.* According to Barloy and Pelhate (1962), *F. solani* knocks down plants in all stages of development, including seedlings. Common aboveground symptoms include partial or systemic wilting. Roots at this stage are red-tinged, rotten, and necrotic. *F. solani* easily invades root wounds created by other organisms, such as nematodes and parasitic plants. Barloy and Pelhate (1962) consider a combination of *F. solani* and broomrape (*Orobanche ramosa* L.) the greatest threat to *Cannabis* cultivation in southern France.

Aboveground symptoms of root rots are hard to tell apart from symptoms of wilt diseases and blight diseases. Wilt diseases are caused by two *Fusarium* species and two *Verticillium* species. The most common cause is *F. oxysporum* f.sp. *cannabis* (Noviello and Snyder, 1962). Wilt symptoms begin with an upward curling of chlorotic leaf tips. Wilted leaves dry to a yellow-tan color and hang on plants without falling off. Cutting into wilted stems reveals a reddish-brown discoloration of xylem tissue. Verticillium wilt is caused by *V. dahliae* (Noviello, 1957) and less commonly *V. alboatrum* (Gzebenyuk, 1984).

Blights are caused by several fungi. *Sclerotium rolfsii* causes southern blight (Ferri, 1961a). Curzi called the fungus *Corticium rolfsii*. Anthracnose is a blight disease caused by *Colletotrichum coccodes* (=*C. atramentarium*) and *Colletotrichum dematium* (Saccardo, 1882; Cavara, 1889). *Macrophomina phaseolina* causes "charcoal rot" and blights plants approaching maturity (McPartland, 1983). It is sometimes called "premature wilt" (Goidànich, 1955). Splitting blighted stems exposes a pith peppered with numerous small black fungal sclerotia, which easily identifies charcoal rot. Brown blight is caused by several *Alternaria* species (*A. alternata* [=*A. tenuis*], *A. solani, A. longipes,* and *A. cheiranthi).* When this disease arises in flowers it can be confused with gray mold (Agostini, 1927; Ferri, 1961b; McPartland, 1983/1995c).

Twig blight is caused by two fungi. *Dendrophoma marconii* was discovered in Pavia, where it is known as *nebbia* (Cavara, 1888). *Botryosphaeria marconii* was discovered by Lyster Hoxie Dewey and described by Charles and Jenkins (1914). Many authors equate *Dendrophoma marconii* with *Botryosphaeria marconii,* but they are different species (McPartland, 1995b/1996a). Twig blight is a serious disease in Europe (Ferraris, 1935; Petri, 1942; Ghillini, 1951). Tips of young branches show wilting symptoms first (thus "twig blight"). Within two weeks the entire plant wilts and dies.

Many diseases are limited to leaf tissue. Who needs leaves, you may think. Of course, the yield of fiber, flowers, and seeds depends on photosynthesis in leaves. So leaf disease decreases crop yield in a highly correlated regression.

The most common leaf disease is yellow leaf spot, caused by two fungi—*Septoria cannabis* and *Septoria neocannabina* (McPartland, 1995d). Yellow leaf spots dry out and fragment, leaving ragged holes in old leaves. In severe infections the leaves curl, wither, and fall off prematurely, defoliating the lower part of the plant (Ferraris, 1935). Spots may also arise on stems and cotyledons of seedlings (Ferri, 1959b). Unlike *Botrytis,* the *Septoria* species do not seem to be seed-transmitted (Ferri, 1961b). Ferraris and Massa (1912) cite

Leptosphaeria cannabina as a "probable" teleomorph of *S. cannabis*. Evidence suggests they erred (McPartland, 1995a/1995d). Brown leaf spot is caused by at least eight different *Phoma* and *Ascochyta* species, some producing *Didymella* ascospores (McPartland, 1995c). The disease is ubiquitous, frequently spreads to stems, and often arises in conjunction with yellow leaf spot. Olive leaf spot is caused by two species, *Pseudocercospora cannabina* and *Cercospora cannabis*. The two fungi occur around the world but rarely seem to cause damage in Europe (McPartland, 1995e). *P. cannabina* seems to be more virulent than *C. cannabis*.

Downy mildew is also caused by two species, *Pseudoperonospora cannabina* (described by Curzi at Pavia) and, rarely, *Pseudoperonospora humuli*. A world map outlining the range of hemp downy mildew, published by the Commonwealth Mycological Institute (1989), is outdated—the disease now occurs on every continent except Antarctica (McPartland and Cubeta, 1997).

Stemphylium leaf spot is sometimes called "brown fleck disease," and frequently spreads to branches and stalks. It is usually caused by *Stemphylium botryosum*, a fungus known by many other names (e.g., *Stemphylium cannabinum, Macrosporium cannabinum, Thyrospora cannabis,* and *Pleospora tarda*—see McPartland, 1995a/1995e for a discussion).

Powdery mildew is caused by two fungi, *Sphaerotheca macularis* and *Leveillula taurica* (McPartland, 1983; McPartland and Cubeta, 1997). This disease is easily confused with pink mildew, caused by *Trichothecium roseum* (Ciferri, 1941; Ghillini, 1951; Ferri, 1961b). The color of these diseases is not always consistent, so the fungi must be examined under a microscope to ascertain their identity.

Other notable leaf diseases include white leaf spot caused by *Phomopsis ganjae* (McPartland, 1984), and black mildew caused by *Schiffnerula cannabis* (McPartland and Hughes, 1995). Four leaf rust fungi are reported from hemp, but each has only been found once (McPartland, 1995a). Pepper spot is caused by *Leptosphaerulina trifolii*, a fungus with a wide host range that results in a large synonymy—including the taxon *Pleosphaerulina cannabina* (McPartland, 1995e). *Curvularia lunata* and *Curvularia cymbopogonis* sometimes expand from leaf spots to blight diseases (McPartland and Cubeta, 1997).

OTHER DISEASES

Nematodes can be a serious problem in *Cannabis*. They usually infest roots. The worst offenders are root knot nematodes. Aboveground symptoms of root knot nematodes are nonspecific—stunting, chlorosis resembling nitrogen deficiency, and midday wilting with nightly recovery. Belowground symptoms are more distinctive —root galls and adventitious rootlets. Galls may coalesce to form "root knots." The most damaging species on hemp is the southern root knot nematode *(Meloidogyne incognita)*, followed by the northern root knot nematode *(Meloidogyne hapla)*, and the Java root knot nematode *(Meloidogyne javanica)*. Nematode wounds provide portholes for *Fusarium solani* and other root rot fungi (McPartland, 1996a).

Cyst nematodes have a similar life cycle to root knot nematodes, except the females swell into hard, lemon-shaped cysts, which become full of eggs. The sugar beet cyst nematode *(Heterodera schachtii)* and the hops cyst nematode *(Heterodera humuli)* have been cited on hemp.

The stem nematode *(Ditylenchus dipsaci)* is different; it migrates up the plant and lives in parts aboveground. Branches and stems become twisted and distorted, with shortened internodes producing stunted plants. The nematode attacks several hundred plant species, and is a serious pest of hemp (Ferraris, 1915/1935; Mezzetti, 1951; Goidànich, 1955). Other nematodes are rarely reported on hemp, including the needle nematode *(Paralongidorus maximus)*, root lesion nematode *(Pratylenchus penetrans)*, spiral nematodes *(Heliocotylenchus* and *Scutellonema* spp.) and reniform nematodes *(Rotylenchulus* spp.).

Bacteria, MLOs, and viruses cause diseases. According to Ferri, (1957a/1957b) and Goidànich and Ferri (1959), several pathovars of *Pseudomonas syringae* cause a variety of hemp diseases, such as bacterial blight (caused by *P. syringae* pv. *cannabina*), striatura ulcerosa (by *P. syringae* pv. *mori*), and wildfire (by *P. syringae* pv. *tabaci*). *Agrobacterium tumefaciens* causes crown gall. *Xanthomonas campestris* pv. *cannabis* causes a leaf spot and blight. Phatak et al. (1975) described an MLO (mycoplasma-like organism) infesting *Cannabis*. Five viruses cause problems—the Hemp streak virus (HSV), Alfalfa mosaic virus (AMV), Cucumber mosaic virus (CMV), Arabis mosaic virus (ArMV), and Hemp mosaic virus (HMV).

Other plants may act as parasites. Broomrape (*Orobanche* spp.) is a root parasite. Dodder (*Cuscuta* spp.) arises as conspicuous tangles of glabrous yellow filaments, twining themselves around stems and branches. Robust specimens girdle branches and pull down hosts (Ferraris, 1935; Ciferri, 1941).

Nutrients cause disease if they are not balanced correctly. Deficiencies of nitrogen, potassium, and phosphorus are the most common. Generally, deficiency symptoms of mobile nutrients (N, P, K, Mg, B, Mb) begin in large leaves at the bottom of plants. Symptoms from less mobile nutrients (Mn, Zn, Ca, S, Fe, Cu) usually begin in young leaves near the top. Nutrient deficiencies are easily corrected if they are properly diagnosed.

Many air pollutants are injurious to plants. Sulfur dioxide causes interveinal leaf chlorosis; hydrogen fluoride vapors cause complete chlorosis between veins and along leaf margins (Goidànich, 1959). Injury from air pollution peaks during daylight hours in warm, humid conditions. "Acid rain" doubles the damage by acidifying soil and creating various nutrient deficiencies. Lastly, genetic diseases may arise in hemp (Crescini, 1956).

This survey is broad but shallow. For more depth regarding individual diseases or pests, see the reviews by McPartland (1996a/1996b), or the booklet by nova-Institute (Gutberlet and Karus, 1995).

CONTROL OF DISEASES AND PESTS

Much of the literature regarding hemp diseases and pests is more than fifty years old. These publications are often antiquated, repetitive, or of only local interest. Furthermore, they date to the age when DDT was considered a miracle cure. We must find new methods.

Some growers renounce the use of all synthetic pesticides. Instead, they use *biocontrol* to manage their diseases and pests. Biocontrol is not new; the use of biocontrol organisms against hemp pests began in France around 1886 (Lesne, 1920). The popularity of biocontrol declined when DDT became king. But biocontrol, like hemp, is enjoying a resurgence. Old-school hemp agronomists rarely used biocontrol; the illicit *Cannabis* horticulturalists have more experience to share. In the 1970s, cultivators in California initially used biocontrol predators. Predators are usually insects, such as lacewings (e.g., *Chrysoperla carnea),* and lady beetles (e.g., *Hippodamia convergens, Rodolia cardinalis,* and *Stethorus picipes*). By definition, a predator must consume *more than one* pest before reaching its adult stage. Each lady beetle and lacewing eats hundreds of aphids, spider mites, scales, thrips, whitefly larvae, budworm eggs, and other pests. But these predators are general consumers. Unfortunately they eat anything, including other biocontrols and beneficial honeybees. *Selective* predators are preferred, such as predaceous mites (e.g., *Phytoseiulus persimilis, Metaseiulus occidentalis,* and *Neoseiulus californicus*). Predaceous mites only eat spider mites and spider mite eggs. Many new predators are becoming widely available. Some are selective, such as the aphid midge *(Aphidoletes aphidimyza).* Others are general predators, such as pirate bugs *(Orius tristicolor, Orius insidiosus,* etc.).

Parasitoids, in contrast to predators, kill their prey from within. Parasitoids only have to consume *one* individual host to reach their adult stage. Adult parasitoids usually insert eggs into pests. The eggs hatch into larvae, which eat pests alive, leaving vital organs for last. The first widely used parasitoids were whitefly wasps *(Encarsia formosa).* Several species of *Trichogramma* wasps have successfully controlled ECBs and budworms in hemp fields (Tkalich, 1967; Marin, 1979; Khamukov and Kolotilina, 1987; Camprag, Jovanic, and Sekulic, 1996). Many new parasitoids are coming on the market, such as leafminer wasps *(Dacnusua sibirica),* thrips wasps *(Thripobius semileuteus),* and ECB tachnid flies *(Lydella thompsonii).*

Microbial pesticides are microscopic pathogens—bacteria, viruses, protozoans, and nematodes. They are specific pest-killers and usually do not harm beneficial organisms. Using these biocontrols is like using chemical control—they come in a can, you mix the contents with water, then spray onto foliage. Our favorite microbial is *Bacillus thuringiensis* (BT). BT bacteria produce toxins that kill specific insects. Some BT strains kill caterpillars, some kill beetle grubs, and some kill maggots. The *Nuclear polyhedrosis* virus (NPV) infects and kills noctùid caterpillars (e.g., cutworms, armyworms, budworms, and some borers). Unfortunately, BT and NPV must be ingested by pests to work. Thus they work poorly against sucking insects (e.g., aphids, whiteflies, leafhoppers). To control sucking insects we apply fungi, such as *Verticillium lecanii, Metarhizium anisopliae, Aschersonia aleyrodis, Paecilomyces farinosus,* and many

Beauveria and *Entomophthora* species. Fungi do not have to be eaten; they work on contact, infecting insects right through their skin. To kill undergrown pests, some growers drench soil with nematodes (*Heterorhabditis* and *Steinernema* species). These nematodes hunt down soil pests, then enter them via natural openings (the mouth, spiracles, or anus). Once inside pests, the nematodes release bacteria (*Xerorhabdis* species). The bacteria produce protein-destroying enzymes that liquefy the pests' internal organs within twenty-four to forty-eight hours. The bacteria also produce antibiotics which prevent putrification of dead pests, allowing nematodes to reproduce in the cadavers.

Biocontrol has not been as successful against disease organisms. Experimentally, the gray mold fungus *(Botrytis cinerea)* is killed by other fungi, such as *Coniothyrium minitans, Ampelomyces quisqualis,* and several species of *Trichoderma* fungi (Reuveni, 1995). *Trichoderma* species also kill other sclerotial soil fungi—*Rhizoctonia solani, Sclerotium rolfsii,* and, to a lesser degree, *Macrophomina phaseolina* and *Sclerotinia sclerotiorum,* as well as nonsclerotial *Fusarium, Pythium,* and *Colletotrichum* species. *Trichoderma* fungi are mixed into the soil or sprayed onto seeds as a seed treatment. Seed treatment works well against damping-off pathogens. Seeds have been coated with other biocontrols—the actinomycete *Streptomyces griseoviridis,* the bacteria *Pseudomonas fluorescens, Burkholderia cepacia,* and *Bacillus subtilis,* and the mycorrhizal fungus *Glomus intraradices.*

To manage most disease problems, we must rely on *cultural* or *mechanical* control. Cultural controls are usually preventative and consist of ordinary farm practices. These methods alter the landscape and make it less favorable for pathogen and pest survival, such as tilling the soil (postseason plowing), destroying infested crop residues after harvest, maintaining proper moisture levels (both soil and air moisture), and the balanced use of fertilizers. Mechanical methods tend to be curative rather than preventative, such as steam sterilizing nematode-infested soil or pruning away fungus-infected branches. The use of resistant crop varieties, regular crop rotation, and the avoidance of seed-borne infestations are all methods that control diseases as well as pests. Pests can also be mechanically controlled with traps—sticky traps, light and color traps, and synthetic pheromone traps.

Mention of synthetic pheromones begins our final topic—chemical control. As long as *Cannabis* continues to be grown in artificial monoculture, we will continue to need pesticides. Technically there is no such thing as a pesticide. "Pesticide" implies that a chemical selectively kills a pest leaving everything else alone. Pesticides are really *biocides*—they harm many living things. Some pesticides are nastier than others. Some pesticides can be used on industrial crops (e.g., fiber hemp) but should never be used on food and drug crops such as seed oil or marijuana. On food and drug crops we prefer *biorational* pesticides. Biorational pesticides are defined as naturally occurring compounds or synthetic analogues of naturally occurring compounds, such as synthetic pheromones. *Natural* means these compounds occur in nature. It does not mean they are safe or belong in baby food. Natural pesticides are quite toxic, *naturally* toxic.

Many biorational pesticides were popular before DDT was invented. Some are simple minerals, such as Bordeaux mixture, a mix of calcium hydroxide (lime) and copper sulfate in water, used as a foliar fungicide. Diatomaceous earth consists of the microscopic fossils of ancient algae. It is formulated as a talc-like dust and tears microscopic holes in the surface of soft-bodied insects.

Other biorational pesticides are organic, containing carbon. Horticultural oil is derived from animals (fish oils), vegetables (seed oils), or minerals (petroleum). Hort oil suffocates small insects. Insecticidal soap is oil combined with sodium or potassium alkali. It also suffocates many insects (but not their eggs). Many organic pesticides are derived from plants—old favorites include pyrethrum, rotenone, quassia, sabadilla, and nicotine. Neem and ryania are powerful new botanicals.

Synthetic botanicals are considered biorational. Examples include imidacloprid, a chlorinated derivative of nicotine, and the many synthetic pyrethroids. Antibiotics are semisynthetic products derived from bacteria. Abamectin is an antibiotic mix of avermectin B1a and avermectin B1b, produced by *Streptomyces avermitilis*. Abamectin is very potent against russet mites, leafminers, and nematodes.

Growth and reproduction regulators (GRRs) are synthetics that mimic natural compounds produced by pests. There are many kinds (Olkowski, Daar, and Olkowski, 1991). *Juvenoids* are synthetic insect growth hormones that arrest insect growth; examples include diflubenzuron, methoprene, and kinoprene. *Reproductive phero-*

mones are GRRs aerosolized by female insects. Males follow pheromone scents to find females. A drop of synthetic pheromone will lure males onto tanglefoot or into funnel traps. Hemp researchers have used sex pheromones against European corn borers, cutworms and armyworms, budworms and hemp moths (Nagy, 1979). *Alarm pheromones* such as β-farnesene cause some pests to drop to the ground where they can be eaten by predators. Other pest species disperse across the surface of plants. Dispersal increases their exposure to pesticides or biocontrols. *Aggregation pheromones* lure pests into death traps, and *feeding stimulants* encourage them to chow down on poisons. Japanese beetle traps, for instance, contain Japonilure (a pheromone bait) and geraniol (a food attractant), which lure the pests into a cul-de-sac trap. The cul-de-sac contains malathion, which kills them.

Malathion is a completely synthetic poison. Some growers totally refuse to use this category of pesticides. Other growers only use synthetic pesticides in baited traps (such as Japanese beetle traps), which keep the chemicals off plants and keep them containerized for effective disposal. *Seed treatment* (coating seeds with pesticides before they are planted) is another relatively safe application of synthetics—little pesticide is introduced into the enviroment, and the pesticide is long gone by the time crops are harvested. Many growers use synthetics selectively—they use *selective* synthetics (such as pimicarb, which selectively kills aphids) with *selective* timing, applied to *selectively* infested plants (not the entire field). In this way, synthetics knock pest populations down to a level that can be managed by biocontrols, with minimal collateral damage to the environment and us.

Categories of synthetic pesticides include *chlorinated hydrocarbons* (e.g., chloropicrin, dicofol, and pentachloronitrobenzene); *organophosphates* (acephate, chlorothalonil, chlorpyrifos, diazinon, dichlorvos, dimethoate, malathion, methyl parathion, and tolclofos-methyl); *carbamates* (benomyl, carbaryl, maneb, and zineb); *amides* (carboxin, iprodione, vinclozolin); *heterocyclic compounds* (captan); and *miscellaneous compounds* (etridiazole, fenbutatin oxide, methyl bromide, and triforine). These pesticides have been used on either fiber crops (Kryachko et al., 1965; Sandru, 1975; Mishra, 1987; de Meijer et al., 1994; van der Werf, van Geel, and Wijlhuizen, 1995; Ashok, 1995; Dippenaar, du Toit, and Botha-Greef, 1996) or marijuana (Frank and Rosenthal, 1978; Clarke, 1981; Frank, 1988).

REFERENCES

Agostini, A. (1927). "Observazioni informa a due ifomiceti saprofiti dannosi di tessuti di canapa." *Atti della Reale Accademia dei Fisiocritici,* 1(3):25-33.
Agrios, G. (1988). *Plant pathology,* Third edition. New York, NY: Academic Press, 803 pp.
Allen, J. L. (1908). *The reign of law, a tale of the Kentucky hemp fields.* New York, NY: Macmillan, 290 pp.
Angelova, R. (1968). "Characteristics of the bionomics of the hemp flea beetle, *Psylliodes attenuata* Koch." *Rastenievudni Nauki,* 5(8):105-114.
Anonymous. (1940). *Lezioni al Corso di Perfezionamento per la Stima del Tiglio di Canapa.* Roma, Italia: Faenza-Fratelli Lega Editori, 332 pp.
Ashok, K. (1995). Chemical control of root rot of ganja. *Current Research—University of Agricultural Sciences* (Bangalore) 24(6):99-100.
Baloch, G. M., Mushtaque, M., and Ghani, M. A. (1974). "Natural enemies of *Papaver* spp. and *Cannabis sativa*," Annual report, *Commonwealth Institute of Biological Control,* Pakistan station, pp. 56-57.
Barloy, J. and Pelhate, J. (1962). "Premieres observations phytopathologiques relatives aux cultures de chanvre en Anjou." *Annales des Épiphyties,* 13:117-149.
Bes, A. (1978). "Prilog poznavanju izgledga ostecenja i stetnosti konopljinog savija-ca—*Grapholitha delineana* Walk." *Radovi Poljoprivrednog Fakulteta Univerzita u Sarajevu,* 26(29):169-189.
Bestagno-Biga, M. L., Ciferri, K., and Bestagno, G. (1958). "Ordina mento artificiale dolle species del genere *Coniothyrium* Corda." *Sydowia,* 12:258-320.
Camprag, D., Jovanic, M., and Sekulic, R. (1996). "Stetocine konoplje i integralne mere suzbijanja." *Zbornik Radova,* 26/27:55-68.
Cavara, F. (1888). "Appunti di patologia vegetale." *Atti dell' Istituto Botanico di Pavia,* Ser. II, 1:425-426.
Cavara, F. (1889). "Materiaux de mycologie Lombarde." *Revue Mycologique,* 11:173-193.
Ceapoiu, N. (1958). *Cinepa, Studiu monografic.* Bucharest, Romine: Editura Academiei Republicii Populare Romine, 652 pp.
Charles, V. K. and Jenkins, A. E. (1914). "A fungous disease of hemp." *Journal of Agricultural Research,* 3:81-84.
Cherian, M. C. (1932). "Pests of ganja." *Madras Agricultural Journal,* 20:259-265.
Ciferri, R. (1941). *Manuale di patologia vegetale.* Roma, Italia: Societa Editrice Dante Alighieri, 730 pp.
Ciferri, R. and Brizi, A. (1955). *Manuale di patologia vegetale.* Vol. III. Roma, Italia: Societa Editrice Dante Alighieri.
Clarke, R. C. (1981). *Marijuana Botany.* Berkeley, CA:And/Or Press, 197 pp.
Commonwealth Mycological Institute. (1989). "*Pseudoperonospora cannabina* on *Cannabis sativa*," Distribution Maps of Plant Disease No. 478, Edition 2.
Conrad, C. (1994). *Hemp: Lifeline to the Future.* Los Angeles, CA: Creative Xpressions Publications, 314 pp.

Crescini, F. (1956). "La fecondazione incestuosa processo mutageno in *Cannabis sativa* L." *Caryologia (Florentinea)*, 9(1):82-92.

Curzi, M. (1927). "*Coniothyrium cannabinum.*" *Atti dell'Istituto Botanico della Università di Pavia*, Ser. III, (3):206.

Curzi, M. and Barabaini, M. (1927). Fungi aternenses. *Atti dell'Istituto Botanico della Università di Pavia*, Ser. III, (3):147-202.

Dippenaar, M. C., du Toit C. L. N., and Botha-Greeff, M. S. (1996). Response of hemp (*Cannabis sativa* L.) varieties to conditions in Northwest Province, South Africa. *Journal of the International Hemp Association*, 3(2):63-66.

Emchuck, E. M. (1937). "Some data on the injurious entomofauna of the truck farms and orchards of the Desna river region." *Traveau de l'Institut de zoologie et biologié. Académie des sciences d'Ukraine*, 14:279-282.

Ferraris, T. (1915). *I Parassiti Vegetali*. Milano, Italia: Ulrico Hoepli Press, 1032 pp.

Ferraris, T. (1935). *Parassiti vegetali della canapa*. Roma, Italia: Riv Agric, 715 pp.

Ferraris, T. and Massa, C. (1912). "Micromiceti nuovi o rari per la flor micologica Italiana." *Annales Mycologici*, 10:285-302.

Ferri, F. (1957a). "La 'striatura ulcerosa' della canapa." *Informatore Fitopatologica*, 7(14):235-238.

Ferri, F. (1957b). "La 'striatura ulcerosa' flagello della canapa." *Progresso Agricolo*, 3(10):1194-1194.

Ferri, F. (1959a). *Atlante delle avversità della Canapa*. Bologna, Italia: Edizioni Agricole, 51 pp.

Ferri, F. (1959b). "La Septoriosi della canapa." *Annali della sperimentazione agraria* N. S. 13:6, Supplement pg. CLXXXIX-CXCVII.

Ferri, F. (1961a). "Sensibilitá di *Sclerotium rolfii* a vari funghicidi." *Phytopathologie Mediterraneanea*, 3:139-140.

Ferri, F. (1961b). "Microflora dei semi di canapa." *Progresso Agricolo (Bologna)*, 7(3):349-356.

Ferri, F. (1961c). "Le avversità delle piante viste alla lente: Canapa." *Progresso Agricolo (Bologna)*, 7:764-765.

Ferri, F. (1963). "Alterazioni della canapa trasmesse per seme." *Progresso Agricolo (Bologna)*, 9:346-351.

Frank, M. (1988). *Marijuana grower's insider's guide*. Los Angeles, CA: Red Eye Press, 371 pp.

Frank, M., and Rosenthal, E. (1978). *Marijuana grower's guide*. Berkeley, CA: And/Or Press, 330 pp.

Ghillini, C. A. (1951). "I parassiti nemici vegetati della canapa." *Notiz. sulle Malatt. delle Piant*, 15:29-36.

Goidànich, A. (1928). "Contributi alla conoscenza dell'entomofauna della canapa. I. Prospetto generale." *Bollettino del Laboratorio di entomologia del R. Istituto superiore agrario di Bologna*, 1:37-64.

Goidànich, G. (1955). *Malattie crittogamiche della canapa*. Bologna-Naples, Italia: Associazione Produttore Canapa, 21 pp.

Goidànich, G. (1959). *Manual di Patologia Vegetale*. Bologna, Italia: Edizioni Agricole, 713 pp.

Goidànich, G. and Ferri, F. (1959). "La batteriosi della canapa da *Pseudomonas cannabina* Sutic and Dowson var. *italica* Dowson." *Phytopathologische Zeitschrift*, 37:21-32.

Gutberlet, V. and Karus, M. (1995). *Parasitäre Krankheiten und Schädlinge an Hanf (Cannabis sativa)*. Köln, Germany: Nova Institut, 57 pp.

Gzebenyuk, N. V. (1984). "The occurrence of fungi on hemp stems." *Miklogiya i Fitopathologiya*, 18(4):322-326.

Harada, T. (1930). "On the insects injurious to hemp, especially *Rhinocus pericarpius*." *Konchu Sekae [Insect World]*, 34:118-123.

Herer, J. (1991). *The emperor wears no clothes*, revised edition. Van Nuys, CA: HEMP Press, 182 pp.

Khamukov, V. B. and Kolotilina, Z. M. (1987). "We are extending utilization of the biological method." *Zashchita Rasteniî*, 4:30-31.

Kryachko, Z., Ignatenko, M., Markin, A., and Zaets, V. (1965). "Notes on the hemp tortrix." *Zashchita Rasteniî Vredit Bolez*, 5:51-54.

Lesne, P. (1920). Une ancienne invasion du "Botys du millet" (*Pyrausta nubilalis* Hb.) en France *Bulletin de la Société de Pathologie Végétale de France*, 7(1):15-16.

Marin A. N. (1979). "The biomethod—in the field!" *Zashchita Rasteniî*, 11:24.

McPartland, J. M. (1983). "Fungal pathogens of *Cannabis sativa* in Illinois." *Phytopathology*, 72:797.

McPartland, J. M. (1984). "Pathogenicity of *Phomopsis ganjae* on *Cannabis sativa* and the fungistatic effect of cannabinoids produced by the host." *Mycopathologia*, 87:149-153.

McPartland, J. M. (1991). "Common names for diseases of *Cannabis sativa* L." *Plant Disease*, 75:226-227.

McPartland, J. M. (1992). "The *Cannabis* pathogen project: Report of the second five-year plan." *Mycological Society of America Newsletter*, 43(1):43.

McPartland, J. M. (1995a). "*Cannabis* pathogens VIII: Misidentifications appearing in the literature." *Mycotaxon*, 53:407-416.

McPartland, J. M. (1995b). "*Cannabis* pathogens IX: Anamorphs of *Botryosphaeria* species." *Mycotaxon*, 53:417-424.

McPartland, J. M. (1995c). "*Cannabis* pathogens X: *Phoma*, *Ascochyta*, and *Didymella* species." *Mycologia*, 86:870-878.

McPartland, J. M. (1995d). "*Cannabis* pathogens XI: *Septoria* spp. on *Cannabis sativa*, sensu strico." *Sydowia*, 47:44-53.

McPartland, J. M. (1995e). "*Cannabis* pathogens XII: Lumper's row." *Mycotaxon*, 54:273-279.

McPartland, J. M. (1996a). "A review of *Cannabis* diseases." *Journal of the International Hemp Association*, 3(1):19-23.

McPartland, J. M. (1996b). "*Cannabis* pests." *Journal of the International Hemp Association*, 3(2):49, 52-55.

McPartland, J. M. and Cubeta, M. A. (1997). "New species, combinations, host associations, and location records of fungi associated with hemp (*Cannabis sativa* L.)." *Mycological Research*, in press.

McPartland, J. M. and Hughes, S. (1994). "*Cannabis* pathogens VII: A new species, *Schiffnerula cannabis.*" *Mycologia*, 86:867-869.
Meijer de, W. J. M., van der Werf, H. M. G., Mathijssen, E. W. J. M., and van den Brink, P. W. M. (1994). Constraints to dry matter production in fibre hemp (*Cannabis sativa* L.). *European Journal of Agronomy* (accepted for publication).
Mezzetti, A. (1951). "Alcune alterazioni della canapa manifestatesi nella decorsa annata agraria." *Quaderni del Centro di Studi per le Ricerche sulla Lavorazione Coltivazione ed Economia della Canapa (Laboratorio Sperimentale di Patologia Vegetale di Bologna)*, No. 11, p. 18.
Mishra D. (1987). "Damping off of *Cannabis sativa* caused by *Fusarium solani* and its control by seed treatment." *Indian Journal of Mycology and Plant Pathology*, 17(1):100-102.
Nagy, B. (1959). "Kukoricamoly okozta elváltozások és károsítási formák kenderen." *Különlenyomat a Kísérletügyi Közlemények (Növénytermesztés)*, 52(4):49-66.
Nagy, B. (1967). "The hemp moth (*Grapholith sinana* Feld., Lepid.:Tortricidae): A new pest of hemp in Hungary." *Acta Phytopathologica Academiae Scientiarum Hungaricae* 2:291-294.
Nagy, B. (1979). "Different aspects of flight activity of the hemp moth, *Grapholitha delineana* Walk., related to integrated control." *Acta Phytopathologica Academiae Scientiarum Hungaricae*, 14:481-488.
Nagy, B. (1986). "European corn borer: Historical background to the changes of the host plant pattern in the Carpathian basin." *Proceedings of the 14th Symposium of the International Working Group on Ostrinia*, pp. 174-181.
Noviello, C. (1957). "Segnalazione di *Verticillium* sp. su *Cannabis sativa.*" *Ricerche, Osserev e Divulg Fitopatol per la Campania ed il Mezzogiorno*, 13-14: 161-163.
Noviello, C. and Snyder, W. C. (1962). "Fusarium wilt of hemp." *Phytopathology*, 52:1315-1317.
Noviello, C., McCain, A. H., Aloj, B., Scalcione, M., and Marziano, F. (1990). "Lotta biologica contro *Cannabis sativa* mediante l'impiego di *Fusarium oxysporum* f.sp. *Cannabis.*" *Annali della Facolta di Scienze Agrarie della Universita degli Studi di Napoli, Portici*, 24:33-44.
Olkowski, W., Daar, S., and Olkowski, H. (1991). *Common-sense pest control.* Newtown, CT: Taunton Press, 715 pp.
Petri, L. (1942). "Rassegna dei casi fitopatologici osservati nel 1941." *Bollettino della R. Stazione di Patologia Vegetale di Roma, N. S.* 22(1):1-62.
Phatak, H. C., Lundsgaard, T., Verma, V. S., and Singh, S. (1975)."Mycoplasmalike bodies associated with *Cannabis* phyllody." *Phytopathologische Zeitschrift*, 83:281-284.
Pimentel, D., McLaughlin, L., Zepp, A., Lakitan, B., Kraus, T., Kleinman, P., Vancini, F., Roach, W. J., Graap, E., Keeton, W. S., and Selig, G. (1991). "Environmental and economic impacts of reducing U.S. agricultural pesticide use." In *CRC Handbook of Pest Management in Agriculture*, Vol. I., Edition II. Boca Raton, FL: CRC Press, pp. 679-686.

Ragazzi, G. (1954). "Nemici vegetali ed animali della canapa." *Humus*, 10(5):27-29.

Rataj, K. (1957). "Skodlivi cinitele pradnych rostlin." *Prameny Literatury*, 2:1-123.

Reuveni, R. (1995). Novel approaches to integrated pest management. Boca Raton, FL: Lewis Publishers, 369 pp.

Rosenthal, E. (1994). *Hemp today.* Oakland, CA: Quick American Archives, 444 pp.

Rothschild, M., Rowan, M. R., and Fairbairn, J. W. (1977). "Storage of cannabinoids by *Arctia caja* and *Zonocerus elegans* fed on chemically distinct strains of *Cannabis sativa.*" *Nature*, 266:650-651.

Saccardo, P. A. (1882-1925 [-72]). *Sylloge Fungorum omnium hucusque cognitorum*, Volumes 1-26. [reprints 1944, 1967], Padova, Italy.

Sandru, I. (1975). Eficacitatea unor insecticide granulate si emulsionabile in combaterea moliei cinepii *(Grapholitha delineana). Probleme de Protectia Plantelor,* 3(2):137-154.

Schmidt, H. E., and Karl, E. (1970). "Ein beitrag zur analyse der virosen des hanfes unter berücksichtigung der hanfplattlaus als virusvektor." *Zentralblatt für Baktteriologie, Parasitenkunde, Infektionskrankheiten und Hygeine, Abteilung,* 2(125):16-22.

Schweinitz de, L. D. (1832). "Synopsis fungorum in America Boreali media degentium," *Trans Amer Philos Soc, New Series,* 4:141-316.

Smith, G. E. and Haney, A. (1973). "*Grapholitha tristrigana* (Lepidoptera: Torttricidae) on naturalized hemp (*Cannabis sativa* L.) in east-central Illinois." *Transactions Illinois State Academy of Science,* 66:38-41.

Sorauer, P. (1958). *Handbuch der pflanzenkrankheiten,* Band 5. Berlin, Germany: Paul Parey, 402 pp.

Stoetzel, M. B. (1989). *Common names of insects and related organisms.* Washington, DC: Entomological Society of America, 195 pp.

Termorshuizen, A. J. (1991). "Literatuuronderzoek over ziekten bij nieuwe potentiële gewassen." In *IPODLO Rapport No. 91-08,* Wageningen, The Netherlands: Instituut voor Planteziektenkundig Onderzoek, Wageningen, 18 pp.

Tkalich, P. P. (1967). "[The occurrence and use of entomophagous insects against the stem borer on hemp.] Vozdelÿvanie i pervichnaya obrabotka konopli," pp. 143-146. Urozhaï, Kiev. [From *Review of Applied Entomology,* 58:3017.]

Traversi, B. A. (1949). "Estudio inicial sobre una enfermeded del girasol (*Helianthus annuus* L.) en Argentina." *Lilloa,* 21:271-278.

Tremblay, E. (1968). "Observations of hemp weevils. Notes on morphology, biology, and chemical control." *Bollettino del Laboratorío Entomologia Agraría "Filippo Silvestri" Portici,* 26:139-190.

van der Werf, H. M. G., van Geel, W. C. A., and Wijlhuizen, M. (1995). Agronomic research on hemp (*Cannabis sativa* L.) in the Netherlands, 1987-1993. *Journal of the International Hemp Association,* 2:14-17.

Chapter 7

Cannabis Germplasm Resources

Etienne P. M. de Meijer

INTRODUCTION

The term "germplasm" refers to any living plant material that can be used for sexual and/or vegetative propagation. It can comprise entire plants, seeds, pollen, bulbs, tubers, rhizomes, cell and tissue cultures, etc. Fortunately, most germplasm occurs *in situ*, i.e., wild plants propagate themselves in natural vegetations and cultivated forms are maintained by farmers and the plant and seed commerce. Germplasm is stored *ex situ* in so-called genebanks, institutions that focus on cultivated plants and their close wild relatives to avoid loss of potentially useful breeding material and to facilitate the utilization by plant breeders. Loss of breeding material (genetic erosion) comprises two features: the loss of allele combinations and frequencies as occurring in specific populations and, more tragic, the total loss of alleles.

In the case of *Cannabis*, "germplasm" refers generally to seed, but with regard to drug strains, also vegetatively propagated clones are commercialized and maintained for many years. Further, it is possible to store frozen pollen for at least a few years. The maintenance and availability of *Cannabis* germplasm has become opportune due to a renewed interest in *Cannabis* cultivation for a range of products in Western countries. Many recent initiatives however, are hampered by the lack of availability of sufficient seed amounts of

---

The author thanks R. G. van den Berg, R. C. Clarke, and Th. J. L. van Hintum for their useful comments on the manuscript.

suitable cultivars. For instance, many of the presently listed fiber cultivars are not actually commercially available and the minority that can be obtained performs only satisfactorily in a limited range of latitudes. In the European Union the choice is further restricted as only cultivars registered in the Union are eligible for subsidy and hence economical to grow. As a consequence some hemp breeding programs have been started. For these and future breeding activities and several studies directed at botanical, forensic, and pharmacological aspects of *Cannabis*, series of well-documented populations are required which reasonably represent the variation in the entire genus or a specified part of it.

This chapter is first directed at the structure of the *Cannabis* gene pool. It then discusses the occurrence of some circumscribed sections in commerce and in germplasm collections.

THE STRUCTURE OF THE **CANNABIS** *GENE POOL*

The genus *Cannabis* can be considered as one isolated primary gene pool (Harlan and de Wet, 1971). All populations can intercross readily and produce fertile hybrids whereas they are reproductively isolated from other genera (Schultes et al., 1974; Small, 1972). Furthermore, it is generally accepted that there are no morphological discontinuities within *Cannabis*. As a consequence there are no valid biological and morphological grounds to discriminate other species than the single *Cannabis sativa* L. However, as this species is extremely heterogeneous some further classification is needed for an adequate specification of populations as well as for the management of germplasm collections. In the opinion of the author, the subdivision of a gene pool of domesticated plants, their wild ancestors, and their naturalized derivatives should ideally take place in one integrated system according to natural and practical criteria. A natural structure reflects as well as possible the genetic diversity and is applied for the discrimination of groups of higher rank. A practical structure uses criteria associated with the utilization by man. It reflects differences in domestication history and economic features and is applied for more detailed subdivisions of lower rank. Current systems for the a priori grouping of *Cannabis* populations will be discussed below. Classifications based on the experimental evaluation of single characteristics such as the commonly applied chemotype classification for *Cannabis* will not be considered in the context of this chapter.

Taxonomic Classifications

Taxonomic classifications claim to reflect the patterns of genetic diversity fairly well and are hence usually followed to make the first discriminations. Breeding behavior and morphological criteria are generally applied for taxon discrimination, but as stated previously, there are no breeding barriers and no morphological discontinuities within the genus *Cannabis*. Nevertheless some authors discriminate at the species level *C. sativa* L., *C. indica* Lam., and *C. ruderalis* Janischevsky (Schultes et al., 1974; Emboden, 1974, 1981; Anderson, 1980). Despite the fact that it would have made more sense to rank the distinguished taxa at the subspecific level, this classification may be of some help to subdivide the *Cannabis* gene pool in large morphological groups. The diagnostic keys employ the features of plant height, degree of branching, achene form, and leaf form. An unambiguous identification cannot be expected due to the continuous variation and the strong plasticity for the first two traits especially. Many populations are intermediate in form and remain unnamed. A distinction can however be made between typical *indica* and typical *sativa* plants. *Indica* is characterized by wide leaflets and a compact branched habit. *Sativa* has narrow leaflets and a slender habit. This seems the only attempt made so far to discriminate groups of *Cannabis* according to pure natural criteria and regardless of the utilization by man. The weedy form *C. ruderalis* is characterized by low stature and small and marbled achenes with a strongly constricted abscission zone. Vavilov (1926) used the name *C. sativa* var. *spontanea* Vavilov, which is a synonym for this same taxon. The discrimination of *C. ruderalis* has little genetic ground, however, as all populations matching the description of this taxon have descended from cultivated *C. sativa* fiber crops and can be expected to differ only in allele frequencies from these.

Small and Cronquist (1976) recognized only one complex species *C. sativa* L. with two subspecies, ssp. *sativa* (not psychoactive) and the psychoactive ssp. *indica* (Lam.) Small and Cronquist, each of these again comprising two botanical varieties (one under cultivation and one run wild). Within the ssp. *sativa* the varieties were called var. *sativa* and var. *spontanea* Vavilov, and within the ssp. *indica*, var. *indica* (Lam.) Wehmer and var. *kafiristanica* (Vavilov) Small and Cronquist, respectively. This discrimination of subspecies and varieties ignores morphological diversity and is based on the state of domestication, the purpose for which plants are cultivated, and their presumed psychoactivity. Using such criteria, Small and Cronquist designed a system that is not natural at all. They distinguished cultivar groups and their respective naturalized derivatives rather than formal taxa.

It is confusing that Small and Cronquist's subspecies of *C. sativa*, ssp. *sativa* and ssp. *indica* do not overlap with the species *C. sativa* and *C. indica* of Schultes et al. (1974), Emboden (1974), and Anderson (1980), as the latter authors circumscribe these taxa irrespective of the psychoactive potency. In agreement with Small and Cronquist, some authors still use the name *indica* exclusively for psychoactive strains, regardless of the habit of the plants considered (e.g., Zohary and Hopf, 1994). The morphological criteria used by Schultes et al., Emboden, and Anderson are generally applied by growers and breeders of drug strains to subdivide *sativa*s and *indica*s among psychoactive populations (e.g., Cherniak, 1982). With respect to cultivated fiber and seed hemp there is little taxonomic disagreement. Only the name *C. sativa* is applied without further subspecific indication. Further, it is fairly common to indicate naturalized populations descended from fiber crops as *C. ruderalis*, *C. sativa* ssp *ruderalis*, or *C. sativa* var. *spontanea*.

Purpose and State of Domestication

When discussing *Cannabis* germplasm it is inevitable that a practical subdivision be made according to the traditional crop utilization. Domestication of *Cannabis* has been aimed at two groups of strains: fiber hemp and drug hemp, used for production of bark fibers and psychoactive resin, respectively. Seed is at present mostly a by-product from fiber hemp cultivars; however, special landraces for seed production seem to have existed in Europe and West Asia (Bredemann, Schwanitz, and Sengbusch, 1957) and do still occur in certain parts of China (Clarke, personal communication, 1997) and perhaps also in Chile (Forster, 1996). It is likely that the distinct domestication histories have caused significant genetic differences between drug and fiber/seed strains as there is little evidence of conscious hybridization of members of the two groups.

It is also relevant to discriminate the various domestication stages *Cannabis* can have. These can vary from truly wild to naturalized (run wild, weedy), landrace, and breeder's cultivar. Current fiber cultivars comprise open-pollinated selections (either from crossbred origin or not) and F_1 and F_2 population hybrids. Modern drug cultivars have a much narrower genetic basis. They comprise clonally propagated genotypes and F_1 full-sib seed-progenies from clones. Within one group of related strains (e.g., a fiber cultivar, its landrace source, and the weedy derivatives of the previous two), differences in domestication status are associated with large differences in agronomic characteristics, in genetic heterogeneity and in allele frequencies.

Geographical Provenance

If still existing, truly or nearly wild *Cannabis* populations can only occur in the hypothesized center of origin Central Asia (Vavilov, 1926) including the Himalayas (Sharma, 1979). At all other locations *Cannabis* has been introduced by man. Hence, a certain geographical occurrence is mostly secondary. As many migrations took place already hundreds to thousands of years ago (Zohary and Hopf, 1994) there has been ample opportunity for local adaptation through mutation and selection. Despite this long history of local adaptation all traits of *Cannabis* still show a continuous pattern of variation among populations. This is obviously due to frequently occurring introgression as a result of long-distance pollen dispersal.

A subdivision according to geographical provenance is frequently applied for fiber hemp because it has some agronomic significance. Provenance, or more correctly adaptation latitude, is closely associated with the phenological development of populations and is used for classifying fiber strains into ecotypes or maturity groups (Northern Russian hemp, Central Russian hemp, Southern Russian or Mediterranean hemp, and Far Eastern hemp). For landraces with a longer history of geographical isolation such grouping may reflect to some extent genetic diversity (e.g., Yumaguzina, Ishtiryakova, and Gilmanova, 1979; Yumaguzina et al., 1979). For the modern

European and West Asian cultivars, however, this classification informs only on phenological characteristics. Considerable mutual genetic relations exist among these cultivars regardless of their ecotype. A limited number of landraces belonging to the Southern Russian (Mediterranean) and Northern and Central Russian ecotypes and their cross-progenies have directly been the basis of, or have been used as a breeding parent for, each of the present European and West Asian cultivars (de Meijer, 1995). The group of Far Eastern hemp cultivated in China, Korea, and Japan seems genetically more distinct from the previous ones. This assumption is supported by the fact that a significant heterosis effect can only be obtained by crossing European and Far Eastern fiber strains and not by crossing the European strains mutually.

Drug-type landraces are quite adequately specified by their provenance as this often indicates the morphological type, traditional use, and even cannabinoid profile. Strains from Afghanistan and Pakistan belong to the *indica* type and are used for hashish production (Clarke, personal communication, 1997). Because of bulked processing of crops aimed at hashish there has been little individual selection which results in strongly varying cannabinoid ratios within populations. Landraces from other countries (Thailand, South Africa, etc.) resemble more the *sativa* habit and are mostly used for marijuana production (Clarke, personal communication, 1997). Due to the possibility to relate superior smoking quality to certain mother plants and the subsequent utilization of their seed to sow the next year's crop, the marijuana strains have already in the landrace stage been effectively selected for high THC and low CBD content. For modern drug cultivars, provenance provides little information on identity or characteristics. Since they are maintained indoors as clones or repeatedly produced as seed progenies from such clones there is no opportunity to adapt genetically.

An Integrated Approach for the Grouping of Cannabis *Germplasm*

To subdivide a priori the entire *Cannabis* gene pool, the available systems must be integrated as none of them is in itself sufficiently comprehensive and detailed. A scheme to cover the entire gene pool is proposed in Figure 7.1. Although it remains to be verified by DNA and protein studies if the taxonomic/morphological distinction between *sativa* and *indica* strains is indeed genetically more significant than any subdivision on practical grounds, this discrimination is first made here to classify groups of higher rank. Nevertheless it is impossible to agree with Schultes et al. (1974), Emboden (1974, 1981), and Anderson (1980) that the taxa *indica* and *sativa* deserve the species status.

Next, according to the purpose of domestication the *sativa* group can be split into fiber/seed dual purpose strains, strains used for seed production solely and marijuana strains, respectively, whereas the *indica* group consists only of hashish strains. It is not clear under what taxonomic group the truly or nearly wild populations, unaffected by domestication, should be placed. Sharma (1979) observed a considerable morphological and chemotypical variation among such populations, but did not discuss the taxonomic consequences and consistently used the genus name to indicate his subject.

As a subsequent criterion, geographical provenance is applied. The fiber/seed hemp is then divided into four geographical groups. Strains solely cultivated for seed exist only with some certainty in the Far East. The *sativa* marijuana, and *indica* hashish landraces are at this level subdivided according to their respective provenance.

With regard to the lowest subdivision, according to domestication status, it must be noted that the naturalized derivatives of cultivated fiber and/or seed strains are discriminated only as a level of domestication. The commonly used taxonomic indications for such populations (*C. ruderalis* and *C. sativa* var. *spontanea*) are omitted as they have little genetic basis. It is presumed in Figure 7.1 that the *sativa* and *indica* drug landraces also have occasionally reverted to a weedy state although references to such phenomena could not be traced. The several manifestations of cultivars, ranging from open pollinated populations to clones, are not included in Figure 7.1 but should of course be taken into account in the maintenance and study of germplasm.

Little is known on the following sections of the gene pool: all Far Eastern hemp and all naturalized and truly wild populations. The survey below is hence mainly focused on the European and West Asian fiber/seed strains and the drug strains.

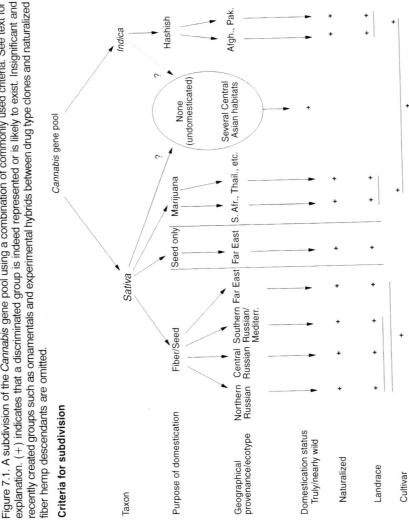

Figure 7.1. A subdivision of the *Cannabis* gene pool using a combination of commonly used criteria. See text for explanation. (+) indicates that a discriminated group is indeed represented or is likely to exist. Insignificant and recently created groups such as ornamentals and experimental hybrids between drug type clones and naturalized fiber hemp descendants are omitted.

Criteria for subdivision

FIBER AND SEED STRAINS

The current commercial strains (printed in boldface) are surveyed by country. When available, information is given on ancestries. Additional details on commercial availability, seed distributors, seed prices, cultivar registration, and some agronomic characteristics have been summarized by de Meijer (1995).

European and West Asian Cultivars

French Cultivars

The current French cultivars can be divided into two groups according to population status. They are either open pollinated populations selected directly from 'Fibrimon' (truly-monoecious cultivars), or they are F_1 and F_2 hybrid cross-progenies of 'Fibrimon' with several dioecious fiber strains (unstable "pseudo-monoecious" cultivars).

'Fibrimon' is a reasonably stable monoecious cross-bred cultivar with high fiber content. It was bred in Germany (Bredemann et al., 1961). Its parental populations were: inbred material obtained from monoecious plants spontaneously occurring in 'Havelländische' or 'Schurigs' hemp which was again a selection of Central Russian origin (Hoffmann, 1961); dioecious selections with very high fiber content from Germany (also retained from Central Russian populations) and dioecious late-flowering landraces from Italy and Turkey. In France, the cultivars **'Fibrimon 21'**, **'Fibrimon 24'**, and **'Fibrimon 56'**, were selected from 'Fibrimon' for diverging dates of maturity. **'Férimon 12'** is again an early maturing selection from 'Fibrimon 21', mainly intended for seed production.

'Fédora 19' is a population hybrid resulting from a cross between female plants of the Russian dioecious 'JUS 9' and monoecious individuals from 'Fibrimon 21', followed by back-crossing of the unisexual female F_1 to monoecious 'Fibrimon 21' plants. The resulting backcross generation (BC_1) is commercialized as first quality seed and consists of 50 percent female and 50 percent monoecious individuals. The open-pollinated seed progeny from such crops is again commercialized as second quality seed. It gives populations with up to 30 percent true male plants. Breeding parent 'JUS 9' is an offspring from a crossing between 'Yuzhnaya Krasnodarskaya' (originally selected from Italian hemp) and dwarf northern Russian hemp.

Likewise, **'Félina 34'** results from a cross between the dioecious parent 'Kompolti', and 'Fibrimon 24', followed by back-crossing to 'Fibrimon 24'.

'Fédrina 74' and **'Futura 77'** both result from a cross between the dioecious parent 'Fibridia' and 'Fibrimon 24' followed by back-crossing to 'Fibrimon 24'. 'Fibridia' is described by Bredemann et al. (1961). It originates from the same German program as 'Fibrimon' and has the same ancestors, except the monoecious 'Schurigs' inbreds.

A new completely THC-free cultivar **'Santhica 23'**, and the nearly THC devoid **'Epsilon 68'** have been registered in France in 1996 (Anonymous, 1996). Their pedigrees are unknown to the author but they are probably based on materials chemotypically described by Fournier et al. (1987).

Hungarian Cultivars

The Hungarian cultivars also comprise open-pollinated populations and F_1 and F_2 hybrids.

'Kompolti' has been selected for high fiber content from 'Fleischmann hemp' or 'F-hemp' which is of Italian origin. The chlorophyll-deficient yellow-stemmed **'Kompolti Sárgaszárú'** was obtained from a cross between a spontaneous yellow-stemmed mutant from Germany (Helle Stengel-Hoffmann, found in the offspring of a cross between Finnish early and Italian late hemp) and 'Kompolti', which was repeatedly back-crossed to 'Kompolti' (Bócsa, 1969).

Hungary is the only country where heterosis breeding of hemp became implemented. This resulted in several F_1 hybrid cultivars. A single cross hybrid cultivar is **'Uniko-B.'** It is the progeny of ('Kompolti' × 'Fibrimon 21') in which the monoecious 'Fibrimon 21' acts as pollen spender. The F_1, being almost unisexual female, is used to produce an F_2, containing ca 30 percent males, which is cultivated for fiber.

'Kompolti Hybrid TC' is a three-way-cross hybrid in which two selections from Chinese landraces, 'Kinai Kétlaki' (dioecious) and 'Kinai Egylaki' (monoecious), and 'Kompolti' are combined. The first step of the crossing ('Kinai dioecious' × 'Kinai monoecious'), in which the monoecious parent acts as pollen spender, gives a unisexual, almost pure female F_1, called 'Kinai Uniszex'. This unisexual progeny can be considered as an analog for male sterile breeding parents. It is subsequently used as seed parent in the crossing ('Kinai Uniszex' × 'Kompolti') which produces the commercial three-way-cross hybrid 'Kompolti Hybrid TC', with a normal 50/50 sex ratio. **'Fibriko'** is the most recent hybrid. It results from a three-way cross in which 'Kinai dioecious' and 'Kinai monoecious' are first crossed to produce the unisexual 'Kinai Uniszex', which is subsequently crossed with the yellow-stemmed pollen spender 'Kompolti Sárgaszárú.' However, 'Fibriko' is not yellow-stemmed, as the normal green stem dominates over yellow.

Polish Cultivars

The two current Polish cultivars are open pollinated monoecious populations. **'Bialobrzeskie'** is the result of a multiple crossing of dioecious and monoecious strains: ((('LKCSD' × 'Kompolti') × 'Bredemann 18') × 'Fibrimon 24'). The dioecious parent 'LKCSD' was selected from 'Havelländische' or 'Schurigs' hemp from Central Russian origin. The dioecious 'Bredemann 18' is a selection from Germany (originally also Central Russian).

The cultivar **Beniko** is a progeny, obtained from the crossing ('Fibrimon 24' × 'Fibrimon 21').

Rumanian Cultivars

The Rumanian cultivars are open pollinated populations.
'Fibramulta 151' is a dioecious selection from the single cross ('ICAR 42-118' × 'Fibridia'). The parent 'ICAR 42-118' is a cross progeny of Italian ('Carmagnola' and Bologna hemp) and Turkish ('Kastamonu') strains (Hoffmann, 1961).

The dioecious **'Lovrin 110'** was selected among family groups from the Bulgarian Silistra landrace ('Silistrenski').

The monoecious **'Secuieni 1'** results from the crossing ('Dneprovskaya 4' × 'Fibrimon') followed by two back-crosses with 'Fibrimon 21' and 'Fibrimon 24', respectively. The Russian dioecious parent 'Dneprovskaya 4' was selected from 'Yuzhnaya Krasnodarskaya' which, again, was obtained from Italian hemp.

Recently the monoecious cultivar **Irene** was released but the breeding history of this cultivar is unknown to the author.

Cultivars from the Ukraine and Russia

Eight open pollinated cultivars are presently cultivated in the central and southern parts of the Ukraine and Russia. Their names generally provide specifications with respect to ecotype (yuzhnaya = southern) and the monoecious character (odnodomnaya). Identical cultivar names, only differing in the added numbers, do not necessarily indicate common ancestry. Due to differences in phonetic translations from Russian, different spelling is sometimes used to indicate the same cultivar.

The dioecious southern type cultivar **Kuban** was obtained by family selection in the progeny from ('Szegedi 9' × 'Krasnodarskaya 56'). The breeding parent 'Szegedi 9' was selected in Hungary from the Tiborszállási landrace. 'Krasnodarskaya 56' is probably a selected cross-progeny from local Caucasian and Italian strains (Hoffmann, 1961).

The monoecious southern cultivar **Dneprovskaya Odnodomnaya 6** is obtained by family selection in the progeny from ('Szegedi 9' × 'Fibrimon 56').

The ancestry of the relatively new dioecious southern cultivar **Zenica** (or 'Shenitsa') is unknown to the author.

The remaining cultivars have a southern phenological pattern but are cultivated at higher latitudes in order to delay flowering and increase stem yield. They are all monoecious.

'Zolotonoshskaya Yuzhnosozrevayushchaya Odnodomnaya 11' (synonyms: 'Zolotonoshskaja 11' and 'Zolotonosha 11'; abbreviated 'USO-11' or 'YUSO-11') is a cross-progeny from 'Dneprovskaya 4', 'YUSO-21', and 'Dneprovskaya Odnodomnaya 6' (Orlov, personal communication, via J. Masura, 1995). The dioecious parent 'Dneprovskaya 4' was selected from 'Yuzhnaya Krasnodarskaya' which again was obtained from Italian hemp. The ancestry of parent 'YUSO-21' is not known.

'Zolotonoshskaya 13' (synonym: 'Zolotonosha 13'; abbreviated 'USO-13' or 'YUSO-13') is a selected cross-progeny from ('YUSO-16' × 'Dneprovskaya Odnodomnaya 6') (Orlov et al., 1987).

'Yuzhnosozrevayushchaya Odnodomnaya 14' (abbreviated: 'YUSO-14' or 'JSO-14') is a further selection from 'YUSO-1', which again is a cross-progeny from ('JUS-6' × 'Odnodomnaya Bernburga').

The dioecious parent 'JUS-6' was selected from ('Yuzhnaya Krasnodarskaya' × 'dwarf Northern Russian' hemp). 'Yuzhnaya Krasnodarskaya' is originally selected from Italian hemp. 'Odnodomnaja Bernburga' was originally produced in Germany in the 1940s under the name 'Bernburger einhäusigen' (Hoffmann, 1961).

'**YUSO-16**' or 'JSO-16' was selected from 'Fibrimon 56'.

'**YUSO-31**' or 'JSO-31' was selected from the crossing ('Glukhovskaja 10' × 'YUSO-1'). The parental population 'Glukhovskaja 10' is a selection from the central Ukrainian Novgorod-Seversk landrace. The ancestry of 'YUSO-1' is described above under 'YUSO-14'.

Apart from these cultivars, the landrace '**Ermakovskaya Mestnaya**' appears to be cultivated on a significant scale in Siberia. It belongs to the Central Russian maturity group. It is not clear whether it is strictly a landrace in the sense that it is maintained only through mass selection by local farmers.

Italian Cultivars

It is doubtful whether this group should be discussed under current cultivars. The EU list of cultivars of agricultural crops and the OECD schemes for the varietal certification of seed moving in international trade together include five open-pollinated Italian cultivars: 'Carmagnola', 'CS', 'Fibranova', 'Eletta Campana', and 'Superfibra'. Due to prohibitive legislation, Italian cultivars have been unavailable for decades. Recently however, 'Carmagnola' and 'Fibranova' have been multiplied again (Grassi, personal communication, 1995).

'**Carmagnola**' is a Northern Italian landrace (Allavena, 1967). 'CS' or 'Carmagnola Selezionata' is dioecious and selected from 'Carmagnola' (Allavena, 1967).

'**Fibranova**' is dioecious and selected from the progeny of 'Bredemann Eletta' × 'Carmagnola' (Allavena, 1961). The parent 'Bredemann Eletta' (or 'Bredemann Elite') is one of the German high fiber selections obtained from Northern and/or Central Russian hemp strains, as were used in the breeding of 'Fibrimon' and 'Bialobrzeskie'.

'**Eletta Campana**' (dioecious) resulted from a cross between the Carmagnola landrace and high fiber strains from German origin, most likely 'Fibridia' or again one of the Bredemann selections.

No information was found on the pedigree of '**Superfibra**'.

Cultivars from the Former Yugoslavia

There is one current open-pollinated Yugoslavian cultivar: **'Novosadska konoplja'**. It is an improved selection from 'Flajsmanova', which is the same as 'Fleischmann hemp' (of Italian origin, see Hungarian cultivars).

German Cultivars

A new open-pollinated, early-maturing German monoecious fiber and seed cultivar, called **'Fasamo'**, has recently become commercially available. It was obtained from a cross-progeny of 'Schurigs' hemp and 'Bernburger einhäusigen', monoecious hemp bred in Bernburg in the 1940s (Loch, personal communication, 1995).

Far Eastern Hemp

Far Eastern hemp consists of landraces only. In Korea, at least eighteen different landraces are cultivated and locally commercialized. Recently a breeding program has been started which is directed at decreasing THC content by means of family selection within landraces and within cross-progenies between low THC European cultivars and local landraces (Moon, personal communication, 1996).

According to Clarke (1995) hemp in China's Shandong province occurs as a real landrace-weed complex as it is not uncommon to use seeds from feral populations for sowing both fiber and seed crops. Shao and Clarke (1996) present several vernacular names for Chinese *Cannabis;* these however refer to distinctive products rather than to genetically distinct populations. Shao and Clarke also give the impression that despite various taxonomic attempts in the recent past, a satisfactory classification of Chinese *Cannabis* is still lacking.

At the beginning of the twentieth century, Chinese landraces together with Northern Italian landraces were used to develop the now extinct Kentucky hemp cultivars which were cultivated until the mid-1950s in the United States (Dewey, 1913, 1927). Ruderal escapes from these cultivars nowadays still persist in the midwestern United States and Canada. At present a selection from Chinese origin is used in Hungary as a parent for heterosis breeding (Bócsa, 1954).

It is not clear if domestic landraces are still cultivated anywhere in Japan.

DRUG STRAINS

Many unnamed *sativa* landraces are still occurring *in situ* in the traditional marijuana-producing countries. Cherniak (1982) mentions Burma, Colombia, Jamaica, Mexico, Nepal, and Thailand as areas with predominant cultivation of *sativa*-type populations. India, the Middle East, Central and South Africa, and the Caribbean can be added to this list (Clarke, personal communication, 1997).

Indica landraces are cultivated for hashish in Afghanistan and Pakistan.

Many population hybrids between *indica* and *sativa* landraces have, either consciously or not, already been made by local farmers (Cherniak, 1982). Since the 1960s, elite clones and families have been obtained from several imported landraces through selective incest breeding in the USA. These, and hybrids between selected clones, form the basis of the modern drug cultivars (Clarke, 1993).

Examples of nonhybrid inbreds are the pure *indica* clones originating from Afghanistan: **'Afghani #1'**, and **'Hindu Kush'**, and a pure *sativa* population from South Africa called **'Durban'**.

Important cross-bred cultivars are Haze, Northern Lights, and Skunk #1. **'Haze'** was developed in California in the 1970s. It is a multihybrid progeny with pure *sativa* landrace ancestors from Colombia, Mexico, Thailand, and Southern India. It was more or less stabilized through incest breeding.

'Northern Lights' was made in Washington State. Its ancestors are one-fourth Thai *(sativa)* and three-fourths Afghani *(indica)* landraces. Northern Lights was far from stable; from the segregating population three female clones with different phenotypes were selected that are currently used for F_1 hybrid seed production: (NL#1, NL#2, NL#5).

'Skunk #1' was bred in California. The ancestors are landraces from Afghanistan *(indica)*, Mexico *(sativa)* and Colombia *(sativa)*. Hybridization took place in the 1970s. Selective incest breeding of the progeny made it a fairly true-breeding cultivar.

The described U.S. cultivars and several cross-combinations made with them are at present more or less legally commercialized in the Netherlands. Private companies sell a total of at least thirty-five different full-sib F_1 seed progenies. In addition, about thirty female clones with similar genetic backgrounds as the seed progenies were made commercially available in recent years. In an attempt to create early cultivars suitable for outdoor cultivation in temperate climates some back-cross hybrids were made approximately ten years ago between drug clones and naturalized fiber hemp derivatives (*C. ruderalis*) from Hungary. However, according to Clarke (personal communication, 1997) this strategy proved to be unsuccessful.

The drug cultivars generally have no official status, i.e., they are not registered and not protected by breeder's rights. An exception is the pharmaceutical drug cultivar **'Medisins'**, a female Skunk clone that was officially registered in the Netherlands in 1996.

OTHER DOMESTICATED CANNABIS

One example of a *Cannabis* cultivar intended for use as an ornamental is the Hungarian cultivar **'Panorama'**. It is a back-cross hybrid between a globe-shaped dwarf mutant that spontaneously occurred in a Lebanese drug strain, and the monoecious fiber cultivar Fibrimon (Bócsa, personal communication, 1990).

CANNABIS *GERMPLASM* EX-SITU

When discussing *ex-situ* collections, a distinction must be made between germplasm collections for use in current breeding and research (working collections) and genebank collections. As opposed to genebank collections, accessions of working collections are generally not freely available and not intended to be eternally maintained. Examples are the working collections at Indiana University (Bloomington, USA, 200 populations for biosystematic studies), at CPRO (Wageningen, Netherlands, 200 original populations and various selections for fiber hemp breeding), at the Ukrainian Research Institute of Bast Crops (Glukhov, Ukraine, an unknown number of original landraces and selections for fiber hemp breeding), and at HortaPharm (Amsterdam, Netherlands, 100 clones and many seed offspring and pollen samples for the breeding of pharmaceutical cultivars).

Only a few genebanks with no direct commercial interest store significant numbers of *Cannabis* seed accessions. Collections occur mainly in connection with recently abandoned or current fiber hemp breeding. By far the largest collection is maintained at the Vavilov Institute (St. Petersburg, Russia). It comprises approximately 500 accessions with a good representation of the entire fiber/seed hemp group including landraces from China (Lemeshev, Rumyantseva, and Clarke, 1994; Kutuzova, Rumyantseva, and Clarke, 1996). The Hungarian genebank (Tapioszele) stores about seventy accessions of local hemp. Smaller collections of up to twenty accessions are preserved in genebanks in Germany, Turkey, and Japan.

Many botanical gardens also maintain one or a few *Cannabis* populations but these are generally not properly specified by relevant attributes such as name, provenance, type of utilization, etc., and are probably all strongly affected by genetic drift.

Cannabis germplasm collections serve mainly practical goals, such as breeding and various types of agronomic, biochemical, and botanical research that generally require well-specified strains with well-defined characteristics. In the author's opinion, germplasm management should hence focus on the preservation of a limited but representative set of populations with their specific characteristics. Landraces and cultivars that fall into disuse are readily lost since *Cannabis* populations are known to show very rapid genetic drift when artificial selection is abandoned or when multiplication takes place under altered conditions. Genetic erosion in the sense of loss of alleles is less likely to happen since due to the frequently occurring introgression, even rare, recessive alleles can be expected to be dispersed diffusely in a wide range of populations over wide areas.

At present *Cannabis* genebank collections are mainly restricted to (formerly) cultivated fiber strains and some of their naturalized derivatives, whereas drug strains and truly wild populations are very poorly represented. This omission is probably largely due to prohibitive legislation with regard to drugs, but it does not make sense from a breeders' point of view. Domesticated drug strains and truly wild populations may be an important source of novel alleles for various future breeding aims, including fiber hemp breeding. As in all other cultivated crop species, it would be a wise decision to establish germplasm collections that cover all domesticated sections of the gene pool and also reasonably represent the wild ancestral populations of the cultivated forms.

REFERENCES

Anonymous (1996). "Nouvelles variétés de lins et chanvre." *Semences et Progrès,* 96:110-111.
Allavena, D. (1961). "Fibranova, nuova varieta di canapa ad alto contenuto di fibra." *Sementi Elette,* 5:34-44.
Allavena, D. (1967). "CS, eine neue Sorte des zweihäusigen Hanfes." *Fibra,* 12: 17-24.
Anderson, L. C. (1980). "Leaf variation among *Cannabis* species from a controlled garden." *Harvard University Botanical Museum Leaflets,* 28:61-69.
Bócsa, I. (1954). "Results of heterosis breeding in hemp." [Hungarian] *Növénytermelés,* 3:301-313.
Bócsa, I. (1969). "Die Züchtung einer hellstengeligen, südlichen Hanfsorte." *Zeitschrift für Pflanzenzüchtung,* 62:231-240.
Bredemann, G., Garber, K., Huhnke, W., and von Sengbusch, R. (1961). "Die Züchtung von monözischen und diözischen, faserertragreichen Hanfsorten 'Fibrimon und Fibridia." *Zeitschrift für Pflanzenzüchtung,* 46:235-245.
Bredemann, G., Schwanitz, Fr., and von Sengbusch, R. (1957). "Aufgaben und Möglichkeiten der modernen Hanfzüchtung mit besonderer Berücksichtigung des Problems der Züchtung haschischarmer oder-freier Hanfsorten." *Technical Bulletin, Max Planck Institut für Züchtungsforschung, Hamburg-Volksdorf,* 4:1-15.
Cherniak, L. (1982). "The great books of *Cannabis.*" Vol. I: Book II. Oakland, CA: Cherniak/Damele Publishing Co., p. 207.
Clarke, R. C. (1993). "Indoor *Cannabis* breeding." In J. Cervantes (Ed.), *Indoor Marijuana Horticulture,* pp. 250-274, Portland, OR: Van Patten Publishing.
Clarke, R. C. (1995). "Hemp (*Cannabis sativa* L.) cultivation in the Tai'an district of Shandong Province, Peoples Republic of China." *Journal of the International Hemp Association,* 2:57-65.
Dewey, L. H. (1913). "Hemp." In *Yearbook of the USDA 1913,* pp. 283-316.
Dewey, L. H. (1927). "Hemp varieties of improved type are result of selection." In *Yearbook of the USDA 1927,* pp. 358-361.
Emboden, W. A. (1974). "*Cannabis*—a polytypic genus." *Economic Botany,* 28: 304-310.
Emboden, W. A. (1981). "The genus *Cannabis* and the correct use of taxonomic categories." *Journal of Psychoactive Drugs,* 13:15-21.
Forster, E. (1996). "History of hemp in Chile." *Journal of the International Hemp Association,* 3:72-77.
Fournier, G., Richez-Dumanois, C., Duvezin, J., Mathieu, J. P., and Paris, M. (1987). "Identification of a new chemotype in *Cannabis sativa:* Cannabigerol-dominant plants, biogenetic, and agronomic prospects." *Planta Medica,* 53: 277-280.
Harlan, J. R. and de Wet., J. M. J. (1971). "Toward a rational classification of cultivated plants." *Taxon,* 20:509-517.

Hoffmann, W. (1961). "Hanf, *Cannabis sativa.*" In H. Kappert and W. Rudorf (Eds.), *Handbuch der Pflanzenzüchtung*, Band V, Paul Parey, Berlin-Hamburg, pp. 204-261.

Kutuzova, S., Rumyantseva, L., and Clarke, R. C. (1996). "Maintenance of *Cannabis* germplasm in the Vavilov Research Institute genebank—1995." *Journal of the International Hemp Association,* 3:10-12.

Lemeshev, N., Rumyantseva, L., and Clarke, R. C. (1994). "Maintenance of *Cannabis* germplasm in the Vavilov Research Institute genebank—1993." *Journal of the International Hemp Association,* 1:1-5.

Meijer de, E. P. M. (1995). "Fibre hemp cultivars: A survey of origin, ancestry, availability, and brief agronomic characteristics." *Journal of the International Hemp Association,* 2:66-73.

Orlov, N. M., Orlova, L. G., Cherevan, A. D., and Lupach, S. M. [Hemp variety Zolotonoshkaya 13] [Russian] Len i Konoplya (1987) No. 3, 43. (Cited from *Field Crop Abstract,* 41:2675.)

Schultes, R. E., Klein, W. M., Plowman, T., and Lockwood, T. E. (1974). "*Cannabis*: An example of taxonomic neglect." *Harvard University Botanical Museum Leaflets,* 23:337-367.

Shao, H. and Clarke, R. C. (1996). "Taxonomic studies of *Cannabis* in China." *Journal of the International Hemp Association,* 3:55-60.

Sharma, G. K. (1979). "A botanical survey of *Cannabis* in the Himalayas." *Journal of Bombay Natural History Society,* 76:17-20.

Small, E. (1972). "Interfertility and chromosomal uniformity in *Cannabis.*" *Canadian Journal of Botany,* 50:1947-1949.

Small, E. and Cronquist, A. (1976). "A practical and natural taxonomy for *Cannabis.*" *Taxonomy,* 25:405-435.

Vavilov, N. I. (1926). "Centers of origin of cultivated plants." In V. F. Dorofeyev, N. I. Vavilov (Eds.), *Origin and geography of cultivated plants.* Cambridge University Press, pp. 22-135.

Yumaguzina, K. A., Ishtiryakova, F. K., and Gilmanova, F. R. (1979). "Immunochemical analysis of reserve proteins in seeds of different hemp ecotypes [Russian]." From *Referativnyi Zhurnal* (1980). (Cited from *Field Crop Abstract,* 34:7472.)

Yumaguzina, K. A., Ishtiryakova, F. K., Mukhametshina, R. T., and Mansurova, R. M. (1979). "Electrophoretic analysis of reserve proteins in seeds of different hemp ecotypes [Russian]." From *Referativnyi Zhurnal* (1980). (Cited from *Field Crop Abstract,* 34:7471.)

Zohary, D. and Hopf, M. (1994). *Domestication of plants in the Old World,* Second edition. Oxford, England: Clarendon Press, pp. 126-127.

Chapter 8

Genetic Improvement: Conventional Approaches

Ivan Bócsa

HISTORICAL REVIEW

As for all plant species, at the beginning of the twentieth century hemp breeding was in advance of genetics. Dewey (1927) bred excellent late-maturing varieties from Chinese landraces (Kymington, Chington, and Arlington), while in one of his varieties he used the Italian landrace Ferrara as one of the parents, as indicated in its name (Ferramington). Dewey can be regarded as the first hemp breeder. In wide spacing his varieties reached a height of as much as 6 m. Unfortunately, due to the hemp production prohibition introduced in the United States, his varieties have been lost.

In Europe, in the first quarter of this century, the excellent German methodologist and breeder Bredemann began a selection for male plants of dioecious hemp in 1924. This was based on *in vivo* determinations of their fiber contents, and led to a considerable acceleration of the improvement of fiber content. In Hungary, Fleischmann (1931, 1934) bred varieties in the 1920s, which were also grown in neighboring countries. His basic stock originated from Italy. In Germany, Heuser (1927) pioneered the work on the anatomical and agronomic questions of hemp fiber content. Schurig (cit. Hoffmann, 1961) developed a variety bearing his name from middle Russian hemp, and this was grown in Germany for several decades (Havelländischer Hanf). Hemp breeding started relatively late in Russia (Vedenski, 1929), but soon a new line of research was initiated: Grishko (1935) and Grishko and Malusha (1935) bred monoecious and simultaneously ripening dioecious hemp varieties. Grishko was the first to realize that all dioecious stands contain monoecious or intersexual types from which monoecious hemp can be bred.

Research on hemp cytology was begun relatively early: Hirata (1928) not only established the $2n = 20$ chromosome number, but also demonstrated the presence of the heterochromosomes. It became clear at an early date that cells with $2n = 40$ chromosomes also occur quite frequently in various root tissues due to endomitosis (Breslavets, 1926) (Postma, 1939 and Mudra, 1934, cit. Hoffmann, 1961). Tetraploid forms were also developed by 1939 (Warmke and Blakeslee, 1939; Ribin, 1939; Hoffmann, 1953; cit. Hoffmann, 1961) and were found to be completely fertile, with larger seeds and seed yields, but their fibers were coarser, so their production was of no economic interest.

The question of sex was the subject of research by McPhee as early as 1925. By means of photoperiodic treatment, it proved possible to develop *subdioicus* forms, i.e., to induce the development of a small number of male flowers on female plants. If these plants were self-fertilized, exclusively female progeny were obtained. In this way, confirmation was given for hemp of the classic experiment carried out by Correns (1907) in *Bryonia,* showing that sex is inherited homo-heterogametically according to Mendel's laws. Later, Grishko (1935) and Grishko and Malusha (1935) described and isolated the first monoecious forms, from which varieties were developed. These authors also described the first simultaneously ripening hemp type, in which there was no difference in the vegetation period between male and female plants.

Work on hemp inbreeding was also begun at an early date (Fruwirth, 1922; Crescini, 1956), as was the study of the inheritance of various properties (hypocotyl color) (Crescini, 1953), in the course of which chiefly morphological mutations were described.

The inheritance of intersexual forms of hemp, together with heterosis breeding and the development of hemp hybrids with various sexual forms (monoecious × dioecious hybrids), brings us up to the present day. Modern hemp breeding and genetics date from the 1950s, and breeders and geneticists from four countries (the Soviet Union, Germany, Italy, and Hungary) played a leading role in this work. Later, in the 1970s, when the importance of hemp declined, hemp research was discontinued in Germany and Italy. In the 1960s, however, hemp breeding was begun in France, where German varieties were first grown. Since the 1970s French researchers have developed their own varieties, which are now dominant in the hemp-producing countries of the EU.

SEX GENETICS

The first intersexual form was described by Antherieth in 1821, but it was some time after the clarification of the inheritance of sex (proof of Mendelian inheritance by Correns in 1905) that research on the sex genetics of hemp was begun.

There is a general consensus on the sexual inheritance of dioecious hemp, but opinions differ greatly on that of intersexual and monoecious forms.

The sexual chromosomes of dioecious hemp were first described by Hirata (1928), but McPhee (1928) found that the heterochromosomes could not be demonstrated in all types. Hirata (1928) proved that in dioecious hemp the inheritance of sex was based on the XY mechanism. McPhee (1925) was the first to induce male or androgynous flowers on female plants by means of photoperiodic treatment. When female flowers on the same or other plants were pollinated by these flowers, the progeny was exclusively female.

However, numerous examinations on intersexual and monoecious forms have shown that inheritance based on the XY mechanism cannot be said to be of general validity.

Hybridization and cytogenetic studies on sex and habit, and on the inheritance of intersexual forms were carried out by Grishko, Levchenko, and Seletski (1937); Neuer and von Sengbusch (1943); Hoffmann (1947, 1952); von Sengbusch (1942); and Köhler (1958, 1961). Although light was thrown on many aspects and a number of results were achieved, this problem is far from being fully solved.

Grishko, Levchenko, and Seletski (1937) suggested that, in monoecious stands, feminized males arose at the expense of purely female forms, and masculinized females at the expense of purely male forms. This means that habit was inherited independently of sex.

In 1943 Neuer and von Sengbusch (1943) repeated McPhee's experiments using their own monoecious hemp, which was practically constant. In the course of their work, aimed at clarifying the inheritance of various intersexual forms of monoecious hemp, crosses involving numerous combinations were carried out and the dioecious ♀ × monoecious intersexual ♂ combinations were found to give 94 percent purely female forms when averaged over approximately 800 plants. The proportion of dioecious normal male forms was around 1 percent, while the remainder were various monoecious forms. The authors stated, obviously based on the results achieved for *Bryonia*, that no dioecious male forms should have been obtained from this cross, and they explained the fact that such forms were nevertheless observed by spontaneous contamination with airborne dioecious pollen. Later, von Sengbusch (1952) added that the habit and sex of various intersexual forms were also determined by a number of allelomorphic sexual manifestation genes. By contrast, Hoffmann (1947, 1952, 1961) interpreted the inheritance of monoecious character as the effect of polygenes on sex and habit, leading to the development of feminized and masculinized series. These genes are present not only in the sexual chromosomes, but also in the autosomes. The 0 to 6 percent ratio of male plants obtained when von Sengbusch's crosses were repeated, and the extremely varied ratio of monoecious forms, were considered to be the result of multigenic inheritance and quite definitely not that of spontaneous open pollination. Hoffmann carried out the crosses using female plants from various dioecious varieties and found that the ratio of purely female plants in the F_1 ranged from 57 to 96 percent. The approximate 1 percent of pure males could not have arisen due to spontaneous open pollination, but must be the result of polygenic inheritance. Neuer and von Sengbusch (1943) attributed a greater role to habit in the inheritance and determination of sex. They suggested that all the forms in the female habit series could be regarded genotypically as females and were thus homogametic (XX), while the series with male habit were genotypically male, irrespective of the sex of the flowers, and were thus heterogametic (XY). According to the hypothesis raised by Hoffmann (1947, 1952) the sex of the flowers is more important than habit. It follows from this that the feminized males are XY forms, like normal males, and the masculinized females are completely identical to normal females, thus being XX forms. Hoffmann thought it was very likely that the inheritance of sex depended not only on the gonosomes, but also to a great extent on the autosomes. This was confirmed by cytological studies. It is clear from Hoffmann's investigations that homo- and heterogametic traits were exhibited by 1-3-cross ($♀$, $♀$, $♀$) monoecious plants with female habit, by feminized males and by masculinized females, so there is every reason to conclude that the sexual chromosomes alone, i.e., the XY mechanism, does not provide a satisfactory explanation of sexual inheritance.

Köhler (1958, 1961) carried out research to decide this question. His investigations demonstrated indubitably that the course of inheritance was polygenic. In his opinion monoecism is determined by two intermediary, additive genes acting in the same direction. The monoecious type with female habit bearing the largest number of male flowers *(subgynoicus)* carries two nonallelic, nonlinked genes in the homozygotic state, while monoecious forms with less male character bear at least four monoecism alleles. These can be considered as weak sexual manifestation genes, which have no effect on habit and act independently of each other and of the factors determining the primary and secondary sexual traits of dioecious hemp. Köhler did not give a definite opinion on the practical interpretation of his research results, but there can be no doubt that he shifted the emphasis in the debate toward the polygenic concept.

This almost twenty-year debate on the sexual genetics of hemp contributed enormously to the solution of certain practical hemp-breeding questions by clarifying the course of inheritance in dioecious × monoecious, monoecious × monoecious, and intersexual forms, thus not only facilitating the breeding of monoecious hemp, but also instigating the development of unisexual hemp in the 1960s.

McPhee, Hirata, Neuer, and von Sengbusch, and Hoffmann all carried out their investigations on $♀$ × monoecious $♂$ crosses purely to clarify theoretical and genetic questions, and none of them considered the practical plant breeding potential latent in an F_1 generation consisting largely of purely female types, which could be used as a male-free maternal parent in intervarietal crosses aimed at the development of unisexual hemp. This possibility was first exploited by Bócsa (1959, 1961), who developed the first unisexual hemp, registered under the name UNIKO-B in 1965, which is still in production today.

The hybridization experiments carried out on intersexual and pure sexual forms gave surprising results. Such experiments were carried out by Grishko, Levchenko, and Seletski (1937); Grishko and Malusha (1935); Neuer and von Sengbusch (1943); von Sengbusch (1942); and Hoffmann (1947, 1952), who all reported similar findings.

If female plants from a dioecious stand (homozygotic in regard to sexual traits) were crossed with ideal monoecious plants, the results described above were obtained, i.e., the F_1 consisted exclusively of purely female plants, even when several thousand plants were examined. If these phenotypically normal female plants (which were, however, heterozygotic for sex) were backcrossed to the "ideal" monoecious parent, the F_1 consisted of 50 percent purely female and 50 percent monoecious plants. If, on the other hand, these heterozygotic female plants were crossed with dioecious males, the sexual ratio characteristic of dioecious hemp, i.e., 50 percent female, 50 percent male, was restored. The same was observed if the "ideal" monoecious plants were crossed with dioecious males. Finally, if heterozygotic female plants were crossed with heterozygotic males the progeny generation was 50 percent female, 50 percent male in regard to habit, while only 36 percent were purely female forms in regard to sex, and the other 14 percent were various intersexual plants with female habit.

It can be seen from the above experiments that female habit and sex are dominant characteristics, as is dioecism. The dominance of dioecism and the recessiveness of monoecism means that although monoecious hemp can be developed with relative ease and is a realistic aim, 100 percent constant monoecism will never be achieved, because, due to the dominant nature of dioecism, dioecious males are bound to occur sooner or later and will then multiply from year to year. Monoecious hemp must thus be regarded as an artificial product which cannot be maintained without human intervention. However, the wide-ranging hybridization experiments carried out on intersexual forms are important not only from the theoretical, but also from the practical point of view, as will be discussed in detail later.

The extremely interesting and surprising results obtained in these hybridization experiments again show that the inheritance of the sex and habit of monoecious and intersexual forms cannot be satisfactorily explained by the XY mechanism valid for dioecious hemp.

In the 1980s the Soviet researcher Migal (1986a, 1986b) began new investigations on sex determination in hemp and found that the gene for lax inflorescence (I) was closely linked to the gene for masculinity (M) on the Y chromosome, while compact inflorescence was linked with the gene for femininity (F). The iIFM genotype is male and the iiFF genotype is female. The masculine genes are dominant over the feminine genes. Like Hoffmann (1961), Migal suggests the presence of sexual genes on autosomes as well as on gonosomes. In the inheritance of monoecism he distinguishes between the purely female phenotype segregating from the monoecious population and the dioecious female genotype proved to exist in crossing experiments.

Sex is genetically determined in *Cannabis* but the sex expression can be modified by several factors. Due to its dioecious nature, *Cannabis* has frequently been used as a test model in physiological experiments investigating the role of phytohormones on sex expression (e.g., Galoch, 1978; Mohan Ram and Jaiswal, 1972; Mohan Ram and Sett, 1981; Mohan Ram and Sett, 1982; Nigam, Varkey, and Reuben, 1981). As a more or less unintended result of this research, various chemical treatments have been published that enable selfing of male and female *Cannabis* genotypes as well as the mutual crossing of individuals of the same sex. As far as the author knows, only recently has chemically induced sex-reversal found practical application in *Cannabis* breeding (de Meijer, personal communication, 1997).

IMPROVEMENT IN STEM YIELD

Selection of Landraces, Family Breeding

Dewey (1927) used simple selection to divide nonhomogeneous populations of Chinese landraces into various types and thus developed his famous varieties (Kymington, Chington, etc.). Fleischmann used a similar procedure starting from landraces from the Bologna and Ferrara regions.

Selection for both habit and vegetation period (stem yield) took place by forming maternal families, the flowering of which was characterized by panmixis. Prior to the work of Bredemann, no selection was made for fiber content, or if it was, it involved only maternal lines. Each year maternal plants were selected from these families on the basis of length of vegetation period, height, diameter, weight, and in some cases, seed weight. The length of the vegetation period and the stem yield are known to be in strong positive correlation with each other. In this way the population could be homogenized relatively quickly, but the quality traits (fiber) could hardly be influenced. This type of family breeding led to the development of the Fleischmann variety, Rumanian, Yugoslav, and other varieties from the famous landrace Carmagnola. Using this method no great improvement in stem yield could be achieved; at most it proved possible to stabilize it at a certain level, though Horkay (1977) expressed the opinion that there were greater reserves for selection for stem yield in the Kompolti hemp variety than for selection for fiber content.

Combination (Hybrid, Heterosis) Breeding

It is probable that the first intervarietal crosses were carried out by Dewey (1927) between the hemp varieties Kymington and Ferrara. These were geographically and phylogenetically distant forms and the F_1 possessed extremely good properties, especially in regard to fiber content and fiber quality. Fleischmann (1938) was the first to use photoperiodic treatment when crossing the Kymington variety, obtained from Dewey, with "F" (Fleischmann) hemp, and in wide spacing the F_1 plants reached a height of 5 m, which was a record in Hungary. Neither Dewey nor Fleischmann, however, thought of utilizing the F_1 generation for general cultivation and of exploiting the heterosis effect (hybrid vigor). In Fleischmann's case this was partly because the F_1 matured too late and had coarse fibers, and partly because the production of F_1 seed by photoperiodic treatment could only be carried out on an experimental scale.

The first state-registered F_1 heterosis (hybrid) hemp (B-7 hybrid) was developed by Bócsa (1954) by crossing stands of Kompolti "F" with a mid-late Chinese hemp. Over the average of three years this F_1 hybrid produced 23 percent more stems and 18 percent more fiber than the Kompolti male parent, and a 21 percent greater stem yield and a 31 percent greater fiber yield than the Chinese female parent. In other words, the F_1 hybrid surpassed both parents in yield ("heterobeltiosis,") (Martinek, 1966), a phenomenon not previously described in hemp. Of the three well-known types of heterosis effect, this is a case of adaptative heterosis, manifested in the approximately 40 percent greater stand density, though the stem height was also the same as that of the best parent. With regard to habit (leaf shape, etc.), the Chinese parent proved to be dominant in the F_1 and the resistance of Chinese hemp to flea beetles *(Psylliodes attenuata)* was also inherited. The length of the vegetation period and the fiber percentage were inherited in an intermediate manner. Although 400-500 kg of hybrid seed could be produced per hectare, this nevertheless proved to be too expensive, because fifteen to twenty working days were required for the removal of male plants from the female rows, and this work was physically exhausting. The intervarietal hybrid was thus only marketed for four to five years.

"Heterosis" hemp was mentioned in Soviet literature at a relatively early date. Arynsteyn (1966), for instance, achieved a heterosis effect for stem, fiber, and seed yields by crossing monoecious and dioecious hemps and then backcrossing to the monoecious form; however, he does not mention how this was utilized, or which parent the hybrid surpassed. In a large number of publications Stepanov (1971, 1974a, 1974b, 1974c, 1975a, 1975b, 1977) reported on the "heterosis" effect manifested in crosses between monoecious and dioecious forms, but these F_1s were regarded purely as initial stocks for the development of new monoecious varieties. To this end he carried out single crosses and backcrosses and also made use of inbreeding (in monoecious hemp) and male sterile forms. The "heterosis" effect amounted to 7 to 20 percent for stem yield and 16 to 27 percent for fiber yield compared to the poorer, earlier monoecious parent. This indicates that the vegetation period was inherited in an intermediate manner. Stepanov regarded hybrids (YUS 11) with better stem and seed yields as good initial stocks for the production of new varieties, but he did not think of using the F_1 heterosis effect directly, although he also examined the general and specific combining abilities of many monoecious $\times$ dioecious hybrids (1975b). In most cases he obtained a heterosis effect for seed yield. Migal and Borodina (1988) described male sterile forms in the YUSO-1426 monoecious variety and even determined the gene responsible for male sterility (a single recessive gene), but they did not consider the possibility of using this in hybrid breeding. The highlight in Soviet research was thus on the crossing of monoecious and dioecious forms, but the heterosis effect manifested in the F_1 has never been used in practice, only in the creation of initial stock for the development of new varieties. In a similar manner Rath (1967), in what was then East Germany, reported on a "heterosis" effect between monoecious and dioecious varieties, but these were really only intermediate forms, as in the case of the Soviet hybrids, and not a genuine case of an F_1 heterosis effect. The only yield component which was as good as or better than that of the best parent was seed yield, but here the phenomenon of "unisexuality" modified the sex ratio in favor of female forms, which could have caused a certain superiority in seed yield over the monoecious parent, since Virovets, Stserbanj, and Sitnik (1976) found that in the same maturity group, dioecious female plants produced a greater seed yield than monoecious plants. This is not surprising, as the plants are more robust and all the flowers are female.

In summary, it can be established that, except in Hungary, heterosis research was restricted almost exclusively to hybrids between monoecious and dioecious hemp, and in the vast majority of cases the yield components studied were inherited in an intermediate manner; in other words, this was not a real heterosis effect compared to the best parent, and the authors either did not wish or were unable to utilize the F_1s directly in cultivation.

In the 1970s, Bócsa (1969, 1971, 1995) developed the old, well-known, hybrid B-7 from a Kompolti × Chinese cross using a unisexual maternal variety. For this purpose it was first necessary to develop a unisexual Chinese F_1 which, as a "male sterile" analog, could be used as a male-free maternal variety. The development of the F_1 hybrid thus required only a fraction of the manual labor previously needed to produce hybrid generations. However, whereas the earlier F_1 hybrid was a single cross (SC), this hybrid was a three-way cross (TC). The Chinese unisexual maternal variety was produced by crossing a Chinese monoecious form with a Chinese dioecious variety. This TC hybrid was state registered in 1982 under the name Kompolti hybrid TC, and is still under cultivation in Hungary. The heterosis effect of the TC F_1 is somewhat smaller than that of the hand-produced B-7 (SC), but is still 6 to 10 percent better than the best parent, and has extremely good seed production: the seed yield of the maternal single cross (unisexual F_1) is very high, reaching a value of 1500 kg/ha under favorable conditions. At present, Kompolti hybrid TC is the only hybrid in the world developed in this way.

Inbreeding

The first data on the effect of inbreeding were published by Fruwirth (1922), who found that inbreeding had no effect on stem length, but nevertheless led to a reduction in stem yield. Fleischmann (1934), who had been carrying out inbreeding since 1919, reported a 50 percent reduction in seed yield. According to Bócsa (1956) the seed yield of Fleischmann's inbred line decreased due to a reduction in thousand seed weight, while the stem yield dropped by 32 percent and the number of leaflets from nine to five. At the same time, early male plants, various intersexual forms, and loose-seeded masculinized female forms were found to segregate. Crescini (1956) observed various mutations as the result of inbreeding: accreted leaflets *(pinaditifolia)* and stem fasciation. Tran Van Lai (1985) evaluated the inbreeding effect for all the major agronomic properties of dioecious and monoecious hemp. He proved beyond any doubt that in the course of inbreeding a reduction is observed first, and to the greatest extent, in the seed yield, then in the fiber content, and finally in the stem yield. There is a shortening of the vegetation period and a decline in pollen fertility.

It should be noted that in the case of dioecious hemp it is impossible to carry out inbreeding in the genetic sense, but only paired breeding, where the increase in the "F" value is slow. The F value of the line inbred by Fleischmann was almost 1 (Tran Van Lai, 1985): $F = 1 - (1 - 1/4^{28}) = 0.9997$.

Inbred hemp was used by Stepanov (1975a) to develop hybrids, but a detailed study of this topic was made by Tran Van Lai (1985). It is interesting to note that hybrids of inbred x noninbred dioecious lines were only 7 to 14 percent better for stem yield than the best parent. The additive genetic variance was relatively small, which means that the heterosis effect was the result of epistasis, arising not from the breeding value of the parents, but from a good combination of alleles or genes. This is confirmed by the fact that the reciprocals of these good combinations do not give a heterosis effect even compared to the parental mean.

The situation is quite different for monoecious hemp, the flower structure of which facilitates spontaneous self-fertilization or artificial self-fertilization. Here the increase in the F value takes place relatively quickly, but due to the nature of monoecism it is impossible to produce hybrids, since the maternal lines cannot be mass castrated. This would require the use of male sterile lines. Although such lines have been described (Migal and Borodina, 1988), no data on their use in hybrid development have been found in the literature.

It is worth noting that the self-fertilization ratio of monoecious hemp is some 20 to 25 percent random mating (panmixia) (Horkay, 1986), since the flower structure facilitates self-fertilization. This explains the inbreeding depression observed for monoecious hemp varieties and the lower stem yield compared to dioecious varieties (Bócsa, 1971; Gāucā et al., 1990). This no doubt explains the conclusion drawn by Amaducci (1969), who stated that the replacement of dioecious Italian varieties by monoecious forms is not justified.

BREEDING FOR AN INCREASE IN FIBER CONTENT

As mentioned in the introduction, the first attempt to breed for an increase in fiber content was undoubtedly made by Bredemann. The importance of his work lies in his discovery that both genders were responsible for an increase in fiber content, while he regarded dioecism as definitely beneficial, since he elaborated a technique based on the vitality and regeneration ability of the hemp plant. He observed that the fiber content of male hemp could be determined prior to flowering by splitting the stem lengthwise and determining the fiber content of one half of the stem. Only the best plants were allowed to flower, while the minus variants were destroyed before flowering. With this method he achieved a tripling in the fiber content over a period of thirty years. The determination of this quality trait prior to flowering, thus enabling pollination to be regulated, is only possible for dioecious hemp (Bredemann, 1924, 1937, 1953, Bredemann et al., 1961). After World War II, numerous new research topics grew out of Bredemann's results and the method itself was improved, as it was technically complicated and slow (the support of the tall plants, the binding up of the long cut surface with paper, etc.). Hoffmann (1943), for instance, discovered that the diameter measured halfway up the stem was correlated with the stem weight (r = 0.80) and with the fiber weight (r = 0.78). Crescini (1954) determined a number of important correlations between various parts of the stem and the fiber content, bast content, and the stem sample to be taken during breeding. Jakobey (1953a) was the first to observe that the fiber content (percentage) decreased parallel to an increase in stem weight, which meant that the comparison of mother plants with different stem weights on the basis of fiber percentage could lead to errors. This is an anatomic correlation, previously noted by Heuser (1927), who approached it from the opposite direction: the ratio of the length and diameter of the stem is correlated with the fiber percentage. Jakobey concluded from the correlation he discovered that an evaluational technique should be elaborated to provide an objective basis for comparison between plants with different stem and fiber weights. This technique, which became known as the "normal axis" method and is still used today, facilitated the rapid graphical distinction of male and female plants that broke the normal correlation. Jakobey (1953a) divided the maternal plants into 50 g stem-weight categories and calculated the mean stem weight and the corresponding individual mean fiber percentage for each category. The former were depicted on the Y axis and the latter on the X axis. By joining the means, a normal axis is obtained, representing the population mean. Each individual plant is then placed in this system of coordinates. If the individual values are closer to the Y axis they are plus variants, while if they are closer to the X axis they are minus variants. In this way plants belonging to each weight group can be brought to a common denominator and it is a simple matter to distinguish individuals with greater stem and fiber weights. This makes it possible to select for stem weight, i.e., stem yield, thus preventing a reduction in stem yield due to constant selection for fiber. Using this method Bócsa (1968) gradually increased the fiber content of Kompolti hemp to 2.7 times the original value over the course of twenty-eight years. By means of selection on a maternal line, Sentchenko (1971) achieved a 1.8 times improvement in fiber content over twenty-three years. Dioecism is a great advantage from the point of view of rapid genetic gain. Considerably later Arnoux, Mathieu, and Castiaux (1969) measured a series of correlations between the anatomic factors of hemp. The graph they published (see Table 8.1), without reference to Jakobey, is basically a normal axis for stem weight and fiber for the variety Fibrimon 56, but they were the

Table 8.1. Correlation coefficients of different bast and fiber properties.

FACTORS	CORRELATION COEFFICIENTS (r)
Stem weight—bast weight	+0.97
Stem weight—fiber weight	+0.94
Wood weight—bast weight	+0.96
Bast weight—fiber weight	+0.97
Wood weight—fiber weight	+0.95
Bast content—fiber weight	+0.91
Total correlations to bast content	
Bast content—stem weight	−0.60
Bast content—stem weight	−0.47
Bast content—stem weight	−0.51
Bast content—stem weight	−0.51
Partial correlation by constant stem weight	
Bast content—wood content	−0.70

Source: Arnoux, M., Mathieu, G., and Castiaux, I. (1969). "L'amélioration du chanvre papetier en France." *Annales D'Amélioration des Plantes,* 4:363-389.

first to state that a continual increase in fiber content will lead to a reduction in dry matter yield, i.e., in stem yield, due to the negative correlation.

Jakobey demonstrated the correlation between stem surface and fiber weight ($r = 0.80$) in 1959. The role of stem surface in the development of fiber percentage was first emphasised by Neuer et al. (1946), who indicated that an increase in fiber percentage was conceivable either by increasing fiber weight or by reducing bast weight.

In regard to the technical improvement of Bredemann's method, Jakobey (1953b) took samples from the side branches of branching male hemp for the *in vivo* determination of fiber content, while Bócsa

(1962, unpublished data) used the upper third of the stem, above the so-called GV point (generative-vegetative point) where leaves begin to grown separately instead of in pairs. This is the beginning of the generative phase. With both sampling procedures the fiber percentage exhibited a very close correlation with the fiber content of the main stem or the lower two-thirds of the stem.

The inheritance of fiber content was examined by Horkay (1977) during twenty-one years of selection on Kompolti hemp. Estimations of h^2 using three methods still gave a value of 0.4 after twenty-one years of selection; the decreasing rate of additive genetic variance made up some 20 to 25 percent of the phenotypic variance, which, with a selection differential of 3.0 to 3.5 percent, led to an increase of 0.60 to 0.88 fiber percent in the progeny generation. Since selection for stem yield was not as strict as that for fiber percentage, the variance for this trait was not exhausted to the same extent.

A distinct negative correlation between fiber quality and fiber content was demonstrated by Kerékgyártó and Nagy (1962). Horkay came to the same conclusion: a 1 percent increase in fiber content led to a reduction of two units in the metric fineness (Nm) value. This was confirmed by histological examinations: the higher the total fiber content of the main stem, the greater the proportion of secondary fiber bundles, i.e., the increase in fiber content due to breeding consisted mainly of secondary fibers (Horkay, 1982). The quality of hemp fiber has now ceased to be a breeding aim, since the various processing and refining technologies (cottonization, steam explosion processing, ultrasonic refining, etc.) have made the slow, complicated work of plant breeding superfluous. Nevertheless, it is interesting to note that analyses carried out by Horkay and Bócsa (1996) showed that male hemp had considerably better quality characters than female hemp, and that dioecious hemp was superior in quality to monoecious forms.

Breeding of Monoecious Hemp

Sex and Habit

This question will be reviewed according to Hoffmann (1961). Sexual dimorphism is extremely pronounced in hemp.

There is a close correlation between sex and habit. In the case of the normal dioecious form, female hemp is characterized by robust, intensively branching plants with close-set flowers, a relatively long vegetation period, and consequently luxuriant, long-lived foliage. Male hemp, on the other hand, is the weaker sex, being less robust and less prone to branching. Its flowers form loose clusters, the foliage is less luxuriant, and the vegetation period is some five to seven weeks shorter.

Numerous transitions exist between the female and male habit. These intermediate forms are known in the literature as intersexual forms.

1. Intersexual forms with female habit:
   a. bearing exclusively female flowers = normal dioecious female *(dioicus* female);
   b. bearing female and androgynous flowers = androgynous female *(gynomonoicus)*;
   c. bearing female, androgynous, and male flowers = androgynous monoecious *(trimonoicus)*;
   d. bearing male and female flowers = monoecious *(monoicus)*;
   e. bearing exclusively androgynous flowers = androgynous form belonging to the female habit series *(diclin)*; and
   f. feminized male plant bearing exclusively male flowers *(dioicus* feminized male)

All six intersexual forms are also found for plants with male habit, giving a total of twelve possible intersexual forms. The last type (f) may also produce a small number of female and androgynous flowers toward the end of the vegetation period.

From the point of view of breeding, the series with female habit is of most importance, and of these chiefly the monoecious type (d), since it has good fertility compared to the series with male habit, which produces very few seeds. On the other hand, the latter probably has a much greater fiber content and better fiber quality. Most of the breeders cited above chose monoecious hemp with female habit as their initial stock. In addition to monoecious hemp, the last member of the series (f) can also be used as one component in breeding "simultaneously maturing" hemp (Grishko, Levchenko, and Seletski, 1937; Hoffmann, 1941). The vegetation period of the latter is the same as that of female hemp and it produces large quantities of pollen.

The various members of the intersexual series with female habit, including the monoecious form (d), can be divided into further intermediate forms, differing from each other only in the number and location of male flowers. According to the classification of Neuer and von Sengbusch (1943) monoecious hemp with female habit can be divided into the following groups:

1. Female habit with few female flowers and many male flowers = ♀⚦, 1-cross monoecious form.
2. Female habit with approximately equal numbers of male and female flowers = ♀⚦, 2-cross monoecious form.
3. Female habit where the female flowers are in the slight majority = ♀⚦, 3-cross monoecious form.
4. Female habit with extremely few male flowers and many female flowers = ♀⚦ and ♀⚦, 4- and 5-cross monoecious forms.

The authors in question referred to 2- and 3-cross monoecious forms as "ideal monoecious" plants. Breeders must aim to obtain a stand consisting entirely of such plants.

It is characteristic of the location of the flowers of ideal monoecious plants that the male flowers are always arranged in whorls at the base of primary and secondary branches, while single female flowers are found at the ends of the shoots.

Breeding Methods and Results

The need to breed monoecious hemp first arose in the Soviet Union (Grishko, Levchenko, and Seletski, 1937; Grishko and Malusha, 1935), since in dual-purpose hemp (fiber and seed on the same stand) the male plants begin to break up and become retted while still standing, so half the crop is ruined and, what is more, does not produce any seed. Grishko, Levchenko, and Seletski (1937) considered that initial stock could best be obtained by selecting the monoecious or intersexual plants occurring spontaneously in all dioecious stands. According to Hoffmann (1938) these occur with a frequency of 0.1 percent when averaged over various dioecious varieties.

Neuer and von Sengbusch (1943) reported that due to the extremely low number of monoecious plants, it was extremely difficult to find them, and when they were found, they were usually already in full flower, which meant that they had already been fertilized by dioecious pollen. Due to open pollination the selected monoecious plants reverted to the dioecious form and monoecious types were not observed again until later generations; even after regular selection they only multiplied slowly in the stand.

Selection methods for sex were described by Neuer and von Sengbusch (1943) and Huhnke et al. (1950). These authors elaborated methods for nipping back the female flowers and for castrating the male flowers. Later, these two techniques were combined with the "one year reserve seed" method, also elaborated by these authors.

Complications arose in the course of this work, as it was difficult to recognize various intersexual forms in time. In order to eliminate unsatisfactory forms, all the monoecious plants were allowed to develop until full flowering. When it became obvious which sex they belonged to, the female flowers of suitable plants were nipped back. These soon re-formed, but by then they could only be fertilized by other suitable plants. The disadvantage of this method was that flowers nipped back in this way produced fewer seeds. In order to overcome this problem, the one-year reserve seed method was elaborated, in the course of which elite lines were sown at two intervals so that the flowering of the two stands did not coincide. The earlier sown plants were the control plot, where the sexual distribution of the various lines could be examined in detail, but from which elite plants were not selected, since all the types were allowed to flower. The "A" plots were sown three to four weeks later and all the lines that were found in the control plot to be unsatisfactory were removed before flowering. In this way only plants from the best lines took part in pollination.

If the stock is inbred the above methods are superfluous, as they are lengthy and expensive. Inbreeding stabilizes monoecism to a great extent, which means that selection for sex can be made directly during flowering (Bócsa, 1958).

Naturally the aim is always to develop pure stocks of the two- and three-cross ($\female^{\male}$, $\female^{\male}$) monoecious types described as "ideal monoecious" forms in the classification of Neuer and von Sengbusch (1943). In general, six to eight generations are required from the appearance or selection of the first monoecious forms to the achievement of almost 100 percent ideal monoecism (Hoffmann, 1961). Both Hoffmann (1947) and Grishko, Levchenko, and Seletski (1937) were in favor of the development of simultaneously maturing hemp, which contained large quantities of feminized male hemp plants in addition to the female forms. However, this type did not become popular in practice. Instead, ideal monoecious varieties bred by von Sengbusch by 1948 became widespread under the name Fibrimon. Later, these varieties became the basis for French hemp breeding. In the meantime monoecious hemp varieties were developed by numerous breeders. In the Soviet Union these included Arynsteyn and Loseva (1958); Arynsteyn and Gurzjy (1959); Arynsteyn and Krenikova (1969a,b); Sentchenko (1970); Virovets, Stserbanj, and Sitnik (1976); and Sentchenko, Virovets, and Stserbanj (1977). In Poland, Obara, Mikolajczyk, Maczkiewicz, Jaranowska, and Kozak (cit. Streletsky, 1970). In Rumania Gaucã (de Meijer, 1995).

The selection of monoecious hemp varieties during maintenance and multiplication is a difficult task. As indicated above, by the 1960s French breeders had switched to the development of "hybrid" populations by means of backcrossing the unisexual F_1 generation to monoecious forms. This backcross to the monoecious form can be carried out on a large scale in the field and the BC_1 generation, which consists of 50 percent monoecious and 50 percent purely female forms, is then backcrossed again to the monoecious form (BC_2), which can be easily kept free of dioecious males by means of negative selection. Seed obtained in this way can be used directly for large-scale production. The varieties Felina 34, Fedora 19, and Fedrina 74 consist of such "hybrid" populations (Mathieu, personal communication, 1995). The only true monoecious varieties are Ferimon and Futura, while the cultivation of the old variety Fibrimon has almost ceased. The problems encountered when attempting to maintain the purity of monoecious hemp are illustrated by the method proposed by von Sengbusch (1952), whereby prebasic seed of the best lines is produced on 0.1 ha. These strictly examined plants provide 100 kg seed, leading to a 0.0001 percent of male plants. The number of male plants in each generation will multiply as follows:

Generation	Area (ha)	Seed (t)	Ratio of male plants
1	0.10	0.1	
2	10.00	6.0	1:100,000
3	100.00	50.0	1:10,000
4	500.00	200.0	1:1,000
5	2000.0-		1:100

The male ratio increases linearly with the size of the area and can never be completely eliminated. It must not be forgotten that the evolutional form of hemp is dioecious, while monoecious varieties are an artificial product, incapable of surviving without human intervention. Within a few generations they revert to dioecism if left to themselves. In agronomic experiments on the French hybrid population of c.v. Felina, Höppner, and Menge-Hartmann (1994) found 8 percent male hemp, despite the fact that, according to the standard, it should have contained none at all. According to other investigations (Bócsa, 1989, unpublished data) the flowering of monoecious forms of the one-cross type in the stand is extremely dangerous, as these increase the appearance of dioecious males in the following generations. This fact further complicates both positive and negative selection in monoecious hemp varieties.

Breeding of Unisexual Hemp

The term "unisexual hemp" was introduced by Bócsa (1959) and was later employed by Hoffmann (1961) in his hemp breeding manual. In the present case unisexuality primarily indicates purely female habit, irrespective of the low proportion of intersexual forms, which also exhibit female habit. Bócsa (1959, 1961, 1966, 1967, 1969) used stand free hybridization to repeat the theoretically important experiments of Neuer and von Sengbusch (1943), which, however, were based on the classic *Bryonia dioica* × *Bryonia alba* cross carried out by Correns. Bócsa used four Hungarian dioecious varieties as maternal parents and his own, not entirely homogeneous monoecious variety as pollinator, according to the following scheme (see Table 8.2).

Table 8.2. Sexual ratio and seed productivity of different unisexual hybrids in Kompolt.

Hybrid	Dioecious ♂ %	Dioecious ♀ %	Different types of intersexual plants %	Total seed-producing plants, females and intersexuals	Seed production relative value	N
Kompolti ♀ × Monoecious ♂	7.7	78.6	13.7	92.3	180.2	1441
Fertődi ♀ × Monoecious ♂	8.9	83.4	7.7	91.1	175.5	1470
Tiborszállási ♀ × Monoecious ♂	8.3	68.7	25.0	91.7	156.0	1469
Chinese ♀ × Monoecious ♂	25.6	66.2	8.2	74.4	169.3	1468
Monoecious ♂ (Control)	5.2	0.6	94.2	94.8	148.9	4507
Kompolti ♀ (Control) dioecious	47.6	52.4	—	—	100.0	1471
(chi)^2		970.75 (P = 0.1%)				

Source: Bócsa, I. (1961). "Data on the seed yielding capacity of the unisexual F_1 - Hybrid hemp." *Növénytermelés,* 10:43-50.

Dioecious ♀ × monoecious ♂
↓
F_1 (vast majority ♀ with a few monoecious and even fewer ♂)
↓
F_2 industrial hemp (dioecism is partially restored)

As can be seen from Table 8.2, the female and monoecious (seed-bearing) forms exhibit a high degree of dominance, which can be exploited in seed production. It was clear from the trial crosses that the greater the homogeneity of the monoecious hemp, the greater the ratio of female and monoecious forms in the F_1. It was thus obvious that such a hybrid should be used to improve seed yield potential, since no previous method had ever led to such a great increase in seed yield potential as this drastic change in the sexual ratio. The Kompolti dioecious ♀ × monoecious ♂ combination was called UNIKO-B F_1. In 1959 this already gave an 80 percent higher seed yield than Kompolti dioecious (Table 8.2). Since there was a great hemp shortage in the country at the time, this was of great economic importance. The variety is still cultivated in Hungary under the name UNIKO-B and it has also been registered abroad. The F_1 is grown exclusively for seed, thus exploiting its great seed-yielding potential, while the F_2 generation is sown as industrial hemp for fiber purposes. (Since the F_1 seed must be produced manually, it cannot be produced in great enough quantities for sowing as industrial hemp, so the F_1 seed is only sufficient for the sowing of special seed hemp crops [2 to 5 kg/ha]. The F_2 generation, nevertheless, possesses satisfactory parameters.) It is interesting that the same monoecious variety gave different ratios of intersexual forms with different varieties; in other words, it is possible to speak of combining ability for unisexuality, but not of sexual heterosis, as this term is unknown in genetics. This method was also taken over by the French, but in different combinations, in backcrosses designed to stabilize monoecious hemp, since their five monoecious hemp varieties could not be maintained in pure form. They called these "hybrid populations" and took the scheme published by von Sengbusch (1952) as their basis (this is in fact a BC_2):

Dioecious ♀ (monoecious ♂ × dioecious ♀ F_1)

F_1 =
50 percent ♀ + 50 percent monoecious (BC_1)

Backcross to monoecious (BC_2)

This is easier to handle that the monoecious forms. The varieties Felina, Fedrina, and Fedora are such backcrossed "hybrid populations," and only Ferimon and Futura are true monoecious varieties (Mathieu, verbal communication, 1995). French breeders had already developed such hybrid populations in the 1960s (Arnoux and Mathieu, 1968; Arnoux, Mathieu, and Castiaux, 1969).

Unisexual hemp can also be employed as the maternal parent in heterosis (hybrid) hemp, since it can be regarded as a "male sterile" analog, which means that F_1 seed can be produced without the need for castration (Bócsa, 1966, 1971, 1995) (Bócsa and Karus, 1997; de Meijer, van der Kamp, and Eeuwijk, 1992).

It is interesting that monoecious × dioecious crosses were carried out in the Soviet Union by Nevinnykh (1962) and Arynsteyn (1966), and that, when discussing hybrids, Stepanov (1974c) even used the term "unisexual," introduced by Bócsa (1959, 1961, 1966, 1995) (Bócsa and Karus, 1997), but there is no information to suggest that the unisexual F_1 is grown on large areas for seed production or that such a state registered variety has yet been developed. Stepanov simply considered these hybrids to be good initial stock. This type of cross was also carried out by Arnoux, Mathieu, and Castiaux (1969) up to the BC_1 stage, but they did not report on the use of unisexual hemp in practice, though the F_1 of a monoecious × dioecious cross was backcrossed twice to monoecious hemp.

BREEDING FOR REDUCED THC CONTENT

The psychoactive substance in hemp (hashish, marihuana, etc.), Δ-9-tetrahydrocannabinol (THC), was not identified until 1964 (Mechoulam, 1970). Until then it was simply referred to as cannabinol. THC became an important factor in Europe when, first in France and then in the whole of the EU, the upper limit for the THC content of cultivated (industrial) hemp varieties was set at 0.3 percent. In Russia the laws are even stricter, with an upper limit of 0.1 percent (Virovets, personal communication, 1997), in order to prevent the use of industrial hemp as a drug.

Hoffmann (1961) cited a lecture presented in Versailles in 1956 by von Sengbusch, who mentioned selection designed to free hemp of its hashish content, and who facilitated the quantitative and qualitative determination of "cannabinol" in thousands of plants by simplifying the "beam" and "diazo" tests, on which, however, no detailed information is available. Using this method he found twenty completely "cannabinol"-free plants, but the further fate of these is unknown. The varieties grown in Germany were analyzed for hashish content as early as 1941 (Hitzemann 1941; cit. Hoffmann, 1961). This is the earliest report on the systematic testing of hemp varieties for hashish content.

In the Soviet Union selection for varieties with an extremely low THC content (JUS-19, JUS-22, etc.) was carried out at a relatively early date (Gorshkova, 1977). At present the whole of the Soviet (Ukrainian) variety collection has the lowest THC content in the world.

In the Soviet Union THC contents were reduced not only in middle Russian and monoecious × dioecious hybrids, but also in varieties belonging to the southern geographical race. Nevinnykh, Nimchenko, and Sukhorodo (1986) reported that after ten selection cycles, the THC content of southern hemp lines was eight to ten times lower than that of the initial stock.

Rivoira et al. (1984) examined the location x variety and sexual interactions of five varieties in regard to THC. They found a significant interaction for sex × location × harvesting date, with great variability with respect to THC content. The THC contents of the major varieties cultivated in Europe were reported by Bócsa and Karus (1997); de Meijer, van der Kamp, and Eeuwijk (1992); and de Meijer (1995).

French breeders initiated a program for the reduction of THC content in the 1970s, but no literature has been found on this subject. French varieties are known to have a low THC content (de Meijer, 1995). Similar work is in progress in Poland (Kurhanski and Chrostowska-Kurhanska, 1994), where lines with THC contents of 0.005 to 0.001 percent, or with only traces of THC have been developed.

A method of selection for low THC content was described by Gorshkova, Sentchenko, and Virovets (1988). The cannabinoid content has a slight effect on the shape and color of the glandular hairs of the bracts. Plants with white-headed, short-bodied glandular hairs contain little or no cannabinoid, while those where the glandular hairs are long-bodied, with yellow-brown heads, contain larger quantities of cannabinoid. Plants that are completely lacking in glandular hairs contain no cannabinoids.

Breeding for low THC content was begun in Kompolt (Research Institute of the Gödöllő University of Agricultural Sciences, Hungary) some twelve years ago and the low THC content of the Kompolti hemp variety is well known (de Meijer, 1995).

The *breeding method* applied in Kompolt consists of sampling the bracts of female plants at seed maturity, since these contain the largest quantity of THC and strict selection is the most important aspect of breeding. These samples, taken from hundreds or thousands of plants, are homogenized and then screened by thin layer chromatography. Minus variants, which give little or no color reaction during this analysis, are further analyzed by means of gas chromatography, after which mother plants of Kompolti hemp with a THC content of 0.01 to 0.03 percent and mother plants of Chinese dioecious with 0.04 to 0.1 percent THC are selected for sowing. (Since the Chinese plants have a higher initial THC content, selection cannot be as strict in this case.) The variability in the cannabinoid content of varieties bred for low THC ranges from 0 to 0.5 percent over several hundred mother plants, and averages 0.11 percent.

In crosses made for breeding purposes it is important that both partners should have as low a THC content as possible. This is more characteristic of varieties belonging to the southern geographical race than of middle Russian varieties.

Recent attempts have been made in Italy to find morphological traits linked with low THC content, so that varieties with a low drug content should be visibly distinct from hashish or marihuana hemps (Ranalli et al., 1996). It should be noted that in Europe, Italy is at present the only country where hemp production has been virtually prohibited until such a morphological marker is found. The method employed by the above authors consists of producing mutants using gamma irradiation, though chemical mutagens are also used (EMS). Various morphological mutants have been obtained, chiefly for leaf shape and color. As yet it is not known how these traits are inherited or how they affect the agronomic value of the excellent initial varieties Carmagnola and Fibranova.

Literature on THC discusses almost exclusively the chemical and therapeutic aspects, while papers on genetics and breeding are very few and far between. Thus very little is known about its inheritance. Soroka (1979) reported that it was inherited in a complicated polygenic manner and that high THC content was dominant in the F_1. However, this makes it possible to select for recessive minus variants. In spite of the polygenic inheritance, the h^2 value is relatively high (Bócsa, unpublished data, 1989), so selection is successful, as proved by breeding results throughout Europe.

RESISTANCE BREEDING

Since there are very few fungal diseases or insect pests which cause epidemics or gradation in hemp, there is extremely little literature on this subject.
Bócsa (1954) described the complete resistance of a Chinese hemp variety to hemp fleas *(Psylliodes attenuata)*, which was dominantly inherited in hybrids developed with southern hemp types. These latter types are known to be resistant to *Orobanche* (Sentchenko and Kolyadko, 1971; Kolyadko, 1971), while the middle Russian varieties and hybrids are susceptible, though this can be reduced by crossing and selection. The most susceptible types to European Orobanche are those belonging to the East Asian geographical race (Bócsa, unpublished data), but resistance is dominant in Chinese × southern type F_1 hybrids, so this is not a problem. Resistance to hemp moth *(Ostrinia nubitalis* Hb) was reported by Virovets and Lepskaya (1983): several of their YUSO varieties and a few Yugoslav, Turkish, French, and Italian varieties exhibited resistance. Finally, de Meijer (1993) reported that varieties more or less resistant to the nematode *Meloidogyne hapla* Chitwood, which may cause damage to hemp grown in a monoculture, were to be found among those he tested. In countries with a Western European maritime climate, resistance to *Botrytis cinerea* is extremely important, as this frequently causes lodging and stem breakage in hemp.

BREEDING FOR SEED YIELD AND OIL CONTENT

An increase in seed yield potential has never been a breeding aim, since hemp seeds are used chiefly for sowing. The oil content is only moderate (28 to 31 percent), so hemp cannot be regarded as a true oil-bearing plant. Dioecious hemp has an extremely poor seed yield, as only half the stand produces seed. The cultivation of monoecious hemp represented a great step forward in improving seed yield, while the appearance of unisexual F_1 hemp led to further progress, since this consists of a vast majority of female dioecious plants, which yield more seed than monoecious plants (Virovets, Stserbanj, and Sitnik, 1976).

With the present monoecious and unisexual varieties a potential seed yield of 1200 to 1500 kg/ha can be achieved, which is unlikely to be further improved without impairing other major agronomic traits.

BIOTECHNOLOGICAL ASPECTS

Mandolino et al. (1996) recently began to perform research using the RAPD technique for the determination of genetic diversity. Interesting correlations were found when investigating sex linked markers.

Attempts have been made to produce callus *in vitro* cultures, but as yet, it has not proved possible to induce shoots, only roots.

REFERENCES

Amaducci, M. T. (1969). "Ricerche sulla tecnice colturale delle canape monoiche utilizzate per fabbricazione di carte pregiate." *Sementi Elette*, 15:166-179.
Arynsteyn, A. I. (1963a). "Directed change in the sex of hemp." *Vestn. sel'skohozjajstv. Nauk. N*, 3:131-133.
Arynsteyn, A. I. (1963b). "A kender ivaránák irányított megváltoztatása." *Vestn. Selk. Hoz. Nauki*, 8(3):131-133. (Russian, Hungarian translation)
Arynsteyn, A. I. (1966). Heterosis in hybrids of different sexual form of hemp. *Citol. Genet. Mezved. Sborn*, 2:103-109.
Arynsteyn, A. I. and Gurzjy, E. Sz. (1959). Results of breeding monoecious hemp varieties. Ves. N-i. Inst, Lyub. Kult. Gluchov. Trudi. 183-201.
Arynsteyn, A. I. and Krenikova, G. A. (1969a). "Methods of breeding high-yielding varieties of monocious hemp." *Ref. Zhurnal*, 23-40. (Russian)
Arynsteyn, A. I. and Krenikova, G. A. (1969b). "Methods of improvement of monoecious hemp with high productivity." *Vozd. i pervich. obr. konopli i kenafa*, Gluchov, 23-40.
Arynsteyn, A. I. and Loseva, Z. E. (1958). "Az egylaki kender nemesítése és az egylakiságot elösegítö feltételek." Trudy po Prikl. *Bot. Gen. i Sel*, 31(3):201-210 (Russian, Hungarian translation).
Arnoux, M. and Mathieu, G. (1968). "Sur la production de variétés hybrides de chanvre." *Techn Textilien*, 11:104-107.
Arnoux, M. and Mathieu, G. (1971). "Qualités et techniques de production des nouveaux chanvres hybrides cultivés en France." *Rostnövények-Fiber Crops Proceedings of International Hemp Breeding Conference*, Kompolt, Hungary: 15-22.
Arnoux, M., Mathieu, G., and Castiaux, I. (1969). "L'amélioration du chanvre papetier en France." *Annales D' Amélioration des Plantes*, 4:363-389.
Bócsa, I. (1954). "Results in heterosis breeding of hemp." *Növénytermelés*, 3:301-316 (Hungarian with English summary).
Bócsa, I. (1956). "New phenomenon of inbreeding in hemp." *Növénytermelés*, 7:1-10 (Hungarian with English summary).
Bócsa, I. (1958). "Beiträge zur Züchtung eines ungarischen monözischen Hanfes und zur Kenntnis der Inzuchterschinungen beim Hanf (*Cannabis sativa* L.)." *Z Pflanzenz*, 39:11-34.

Bócsa, I. (1959). "Vorläufiger Bericht über die Herstellung der unisexuellen (von männlichen Pflanzen freien)." *Hanfform Kisérl Közl*, LII/A:135-142 (Hungarian with German summary).
Bócsa, I. (1961). "Data on the seed-yielding capacity of the unisexual F_1—Hybrid hemp." *Növénytermelés*, 10:43-50 (Hungarian with English summary).
Bócsa, I. (1966). "Neue Richtungen und Möglichkeiten in der Züchtung des südlichen Hanfes in Ungarn." *Acta Agric Scandinavica*, 16:292-294.
Bócsa, I. (1967). Züchtung von unisexueller (männchenfreien) Hanfmuttersorte für die Erzeugung von Hanf-Sortenhybriden." *Rostnövények-Fiber Crops*, 3-8 (Hungarian with German summary).
Bócsa, I. (1968). "Die Ergebnisse der Hanfzüchtung auf Fasergehalt in Ungarn." *Rostnövények-Faserpflanzen-Fiber Crops* (Hungarian with German summary).
Bócsa, I. (1969). Möglichkeit der Umänderung der ausländischen einhäusigen Hanfbaues mit Hilfe von neuen Formen des unisexuellen Hanfes." *Rostnövények-Fiber Crops*, 27-32.
Bócsa, I. (1971). "Verwirklichung von besonderen Zielsetzungen in der Hanfzüchtung." *Rostnövények-Fiber Crops Proceedings of International Hemp-Breeding Conference*, Kompolt, Hungary: 29-34.
Bócsa, I. (1995). "Die Hanfzüchtung in Ungarn: Zielsetzungen, Methoden und Ergebnisse." *Symposium "Biorohstoff Hanf,"* Frankfurt a. Main: 200-215.
Bócsa, I. and Karus, M. (1997). "Der Hanf-anbau." *C. F. Müller Verlag-Heidelberg*.
Bredemann, G. (1924). "Beiträge zur Hanfzüchtung II. Anslese faserreicher Männchen zur Befruchtung durch Faser-bestimmung an der lebeden Pflanze vor der Blüte." *Angew Bot*, 6:348-360.
Bredemann, G. (1927). "Beiträge zur Hanfzüchtung III." *Z Pflanzenz*, 12:259-268.
Bredemann, G. (1937). "Züchtung des Hanfes auf Fasergehalt." *Z Pflanzenz*, 12:259-268.
Bredemann, G. (1953). "Verdreifachung des Fasergehalts bei Hanf durch fortgesetzte Männchen und Weibchenauslese." *Mater Veg*, 2:167-187.
Bredemann, G., Garber, W., Huhnke, W., and von Sengbusch, R. (1961). "Die Züchtung von monözischen und diözischen, faser-ertragsreichen Hanfsorten Fibrimon und Dibridia." *Z Pflanzenz*, 46:235-245.
Breslavets, L. (1926). "Polyploide Mitosen bei *Cannabis sativa* L. I." *Ber. Deutsch. Bot. Ges.*, 44:498-502.
Correns, C. (1907). Die Bestimmung und Vererbung des Geschlechtes nach Versuchen mit höheren Pflanzen." *Verh. Ges. Naturforsch. u. Ärzte*, Berlin.
Crescini, F. (1953). "Associazione di fattori genetici (lincage) in *Cannabis sativa* L." *Caryologia*, V:289-296.
Crescini, F. (1954). "Ricerche intorno al miglioramento genetico della canapa da tiglio coltivata in Italia." *Caryologia*, 6(2-3):284-318.
Crescini, F. (1956). "La fecondazione incestuosa processo mutageno in *Cannabis sativa* L." *Caryologia*, IX:82-92.
Dewey, L. H. (1927). "Hemp varieties of improved type are result of selection." *Yearbook of the Department of Agriculture, USA*: 358-361.
Fleischmann, R. (1931). "Hanf-und Flachskultur in Ungarn." *Faserforsch*, 9:143-149.
Fleischmann, R. (1934). "Beiträge zur Hanfzüchtung." *Faserforsch*, 11:156-161.
Fleischmann, R. (1938). "Der Einfluss der Tageslänge auf die Entwicklung von Hanf und Ramie." *Faserforschung*, 13:93-94.

Fleischmann, R. and Bócsa, I. (1951). "Contributions to the improvement of hemp and to the possibility of breeding monoecious varieties" *Agràrtud,* 4:1-3 (Hungarian with English summary).
Fruwirth, C. (1922). "Hanf Handbuch der landwirtschaftlichen Pflanzenzüchtung." *Band III.* Berlin: Paul Parey.
Galoch, E. (1978). The hormonal control of sex differentiation in dioecious plants of hemp *(Cannabis sativa). Acta Societat is Botanicorum Poloniae,* XI, VII: 153-162.
Gauca, C., Trotus, E., Roman, M., Paraschiviou, R., Sim, M., Ursachi, F., and Moisa, F. (1990). "New elements in the technology of seed production in monoic hemp." (Elemente noi in tehnologia producerii de saminta la cinepa monoica). *Analele ICCPT Fundulea* (Rumanian with English summary).
Gorshkova, L. M. (1977). "Prospects of breeding hemp for low content of cannabinoid compounds." *Gent. i Sel. Rast,* 130 (Thesis PhD in Russian).
Gorshkova, L. M., Sentchenko, G. I., and Virovets, V. G. (1988). "Method of evaluating hemp plants for content of cannabinoid compounds." *Referativny Journal 12,* 65:322.
Grishko, N. N. (1935). "Neues in der Hanfselektion. Berichte der allrus." Akad. der Wiss. f. Landw. Ser. 3.4.10-15.
Grishko, N. N., Malusha, K. V. (1935). "Probleme und Richtlinien in der Hanfzüchtung." *Trudy prikl,* 74:61-67 (Russian) (cit: Hoffmann, 1961).
Grishko, N. N., Levchenko, V. I., Seletski, V. I. (1937). "Question of sex in hemp, the production of monoecious forms and of varieties with simultaneous ripening of both sexes." *Vszesoy Nauchno-Issled Inst Konopli,* Moscow, Leningrad. 5:73-108 (Russian with English summary).
Heuser, O. (1927). *Die Hanfpflanze.* Berlin: Springer Verl.
Hirata, K. (1928). "Sex determination in hemp *(Cannabis sativa* L.)." *J Genet,* 19:65-79.
Hitzemann, W. (1941). "Untersuchungen auf Haschisch bei verschiedenen Hanfsorten eigenen Anbaus in Deutschland." *Archiv d Pharmazie,* 279:353-387 (cit. Hoffmann, 1961).
Hoffmann, W. (1938). "Das Geschlechtsproblem des Hanfes in der Züchtung." Z. *Pflanzenz,* 22:453-461.
Hoffmann, W. (1941). "Gleichzeiting reifende Hanf." *Züchter,* 13:277-283.
Hoffmann, W. (1943). "Hanf, *Cannabis sativa* L." In Roemer-Rudorf (Ed.), *Handbuch der Pflanzenzüchtung,* Bd. IV, 314-341.
Hoffmann, W. (1946). "Helle Stengel: Eine wertvolle Mutation des Hanfes *(Cannabis sativa* L.)." *Züchter,* 17/18:56-59.
Hoffmann, W. (1947). "Die Vererbung der Geschlechtsformen des Hanfes *(Cannabis sativa* L.)." *I Der Züchter,* 17/18:257-277.
Hoffmann, W. (1952). "Die Vererbung der Geschlechtsformen des Hanfes *(Cannabis sativa* L.)." *II Der Züchter,* 22:147-158.
Hoffmann, W. (1961). "Hanf." In *Rudorf-Kappert: Handbuch der Pflanzenzüchtung V,* Berlin-Hamburg: Paul Parey, pp. 204-264.
Höppner, F. and Menge-Hartmann, U. (1995). "Cultivation experiments with two fiber hemp varieties." *Journal of the International Hemp Association,* 2(1):18-22.
Horkay, E. (1977). "Investigations on the heredity of fiber content and dry stalk production in Kompolti hemp." *Növénytermelés,* 26:451-459 (Hungarian with English summary).

Horkay, E. (1982). "Primary and secondary fiber-cell ratio as affected by selection for increasing fiber content." *Növénytermelés*, 31:297-301 (Hungarian with English summary).

Horkay, E. (1986). "Establishing the share of self- and cross-fertilization by means of population geneties in a monoecious hemp stand." *Növénytermelés*, 35:177-182 (Hungarian with English summary).

Horkay, E. and Bócsa, I. (1996). Objective basis for evaluation of differences in fiber quality between male, female, and monoecious hemp. *Journal of the International Hemp Association*, 3(2):67-68.

Huhnke, W., Jordan, C., Neuer, H., von Sengbusch, R. (1950). "Grundlagen für die Züchtung eines monözischen." *Hanfes Z Pflanzenz*, 29:55-75.

Jakobey, I. (1953a). "Fiber content value estimation of the hemp stalk by analysis of branches fiber content—(A kenderkòrò rosttartalmi értékének meghatàrozàsa oldalàgai ùtjàn)." *Növénytermelés*, 2:238-247 (Hungarian).

Jakobey, I. (1953b). "The objective evaluation of bast content in fiber crops—(Hàncsrosttartalmù növények objektìv értékelése)." *Növénytermelés*, 2:144-150 (Hungarian).

Kerékgyártó, P. and Nagy, K. (1962). "Industrielle Bewertung des veredelten ungarischen Faserhanfs." *Fibra*, 7(2):61-78.

Köhler, D. (1958). "Zur Vererbung der Monözie beim Hanf." *Z. Vererbungslehre*, 89: 437-447.

Köhler, D. (1961). "Ein Beitrag zur Physiologie und Genetik der Geschlechtsausprägung von *Cannabis sativa* L." *Planta*, 56:150-173.

Kolyadko, I. V. (1971). "Resistance of hemp varieties to Orobanche ramosa." *Len i Konoplya*, 12:27.

Kurhanski, M. and Chrostowska-Kurhanska, G. (1994). "Biological principles for breeding non-narcotic hemp (Przyrodnicze podstawy uzyskania konopi nienarkotycznych)." *Natural Fibers*, XXXVIII:19-26.

Mandolino, G., Faeti, V., Zottini, M., Moschella, A., and Ranalli, P. (1996). "Il miglioramento genetico della canapa: aspetti biotecnologici." *Sementi Elette*, (XLII) 2:57-60.

Martinek, S. F. (1966). "Heterosis, heterobeltiosis, diallel analysis and gene action in crosses of *Triticum aestivum*." *Diss Abstr Ann Arbor*, 26:4153.

Mathieu, G. (1995). Personal communication.

McPhee, H. (1925). "The influence of enviroment on sex in hemp." *Journal of Agricultural Research*, 31:935-943.

Mechoulam, R. (1970). "Marijuana chemistry." *Science*, 168 (3936):1159-1166.

Meijer, de E. P. M. (1993). "Evaluation and verification of resistance to *Meloidogyne hapla Chitwood* in a *Cannabis* germplasm collection." *Euphytica*, 71:49-56.

Meijer, de E. P. M. (1995). "Fiber hemp cultivars. A survey of origin, ancestry, availability and brief agronomic characteristics." *Journal of the International Hemp Association*, 2:66-72.

Meijer, de E. P. M., van der Kamp, H. I., van Eeuwijk, F. A. (1992). "Characterization of *Cannabis* accessions with regard to cannabinoid content in relation to other plant characters." *Euphytica*, 62:187-200.

Migal, N. D. (1986a). "Genetic determination of sex in hemp. II Sexual mutations and the general theory of genotypic control of sex in hemp." *Genetika USSR*, 22:829-837 (Russian with English summary).

Migal, N. D. (1986b). "Genotypic determination of sex in hemp. IV Monoecious hemp." *Genetika USSR:* 2115-2125 (Russian with English summary).

Migal, N. D., Borodina, E. I. (1988). "In heritance and practical use of the intersexual form of male sterility in hemp." *Tsitologya i Genetika,* 22:40-45 (Russian with English summary).

Mohan Ram, II.Y. and Juiswal, V. S. (1972). "Induction of male flowers on female plants of *Cannabis sativa* by gibberellins and its inhibition by abscisic acid." *Plants,* 105:263-266.

Mohan Ram, II.Y. and Sett, R. (1981). "Modification of growth and sex expression in *Cannabis sativa* by aminoethoxyvinylglycine and ethephon." *Z. Pflanzenphysiol.,* 105:165-172.

Mohan Ram, II.Y. and Sett, R. (1982). "Induction of fertile male flowers in genetically female *Cannabis sativa* plants by silver nitrate and silver thiosulphate anionic complex." *Theor. Appl. Genet.,* 62:369-375.

Neuer, H., von Sengbusch, R. (1943). "Die Geschlechtsvererbung bei Hanf und die Züchtung eines monözischen Hanfes." *Züchter,* 15:49-62.

Nevinnykh, V. A. (1962). "Gibridizatsya jushnoy dvoudomnoy konoply." *Agrobiologya,* 2:205-215.

Nevinnikh, V. A., Nimchenko, P. V., Sukhoroda, T. (1986). "Results of work in breeding southern hemp for lack of active cannabinoids." *Selektsya i semen Korm i tekc kultur,*: 45-50 (Russian with English summary).

Nevinnich, V. A., Nimchenko, P. V., Sukhoroda, T. (1986). "Results of work in breeding southern hemp for lack of active cannabinoids." *Selektsya i Semen,* Krasnodar: 45-50 (Russian).

Nigam, R. K., Varkey, M., and Reuben, D. E. (1981). "Streptovariein-induced sex expression in male and female plants of *Cannabis sativa* L."*Ann. Bot.,* 47:169-172.

Ranalli, P., Di Candilo, M., Marino, A., Zottini, M., Fuschi, P., Polsinelli, M., and Casarini, B. (1996). "Induzione di mutanti in *Cannabis sativa* L." *Sementi Elette (XLII),* 2:49-55.

Rath, L. (1967). "Heterosisversuche mit Hanf. Techn." *Textilien,* 5:247-260.

Rivoira, G., Spanu, A., Marras, G. F., Amaducci, M. T., Venturi, G., Benati, R., Basso, F., and Basso, E. V. (1984). "Content of 9-THC in European varieties of *Cannabis sativa.*" *Quaderni di ricerca,* 2:3-19.

von Sengbusch, R. (1942). "Beitrag zum Geschlechtsproblem beim *Cannabis sativa.*" *Z Vererbungsl,* 80:617-618.

von Sengbusch, R. (1952). "Ein weiterer Beitrag zur Vererbung des Geschlechts bei Hanf als Grundlage für die Züchtung eines monözischen Hanfes." *Z F Pflanzenz,* 31:319-338.

Sentchenko, G. I. (1970). "A contribution to the discussion on problems of introducing monoecious hemp into agriculture." *Len i Konoplya* (Flax and Hemp), 5:30-32.

Sentchenko, G. I. (1971). "Ergebnisse der Hanfzüchtung auf Fasergehalt." *Rostnövények-Fiber Crops Proceedings of International Hemp Breeding Conference,* Kompolt, Hungary: 137-144.

Sentchenko, G. I. and Kolyadko, I. V. (1971). "The resistance of hemp hybrids to Orobanche ramosa." *Trudy VNII. Lub. Kult.,* 34:46-50.

Sentchenko, G. I., Virovets, V. G., and Stserbanj, I. I. (1977). "Intervarietal hibridization: The basic method of the monoecious hemp improvement." *Biol. vozd. i pervich. obrob. kon. i kenafa,* Gluchov, 40:3-12.

Sokora, V. P. (1979). "Breeding of monoecious hemp with minimal THC-content." *Szelek Szemen*, 4:11-12.
Stepanov, G. S. (1971). "Heterosis in hemp." *Len i Konoplya* (Flax and Hemp), 16:33-34 (Russian).
Stepanov, G. S (1974a). "The manifestation of heterosis in unisexual hemp hybrids." *Tsitologya i Genetika*, 8:111-114 (Russian with English summary).
Stepanov, G. S. (1974b). "Particulars of choosing hemp varieties in breeding for heterosis." *Tsitologya i Genetika*, 8:441-444 (Russian with English summary).
Stepanov, G. S. (1974c). "Heterosis in unisexual hemp hybrids." In *Referativny Journal*, 12.55.237.
Stepanov, G.S. (1975a). "The use of the inbreeding method in breeding hemp." In *Referativny Journal*, 11.55.236 (Russian).
Stepanov, G.S. (1975b). "Evaluation of combining ability in hemp varieties in breeding for heterosis." *Tsitologya i Genetika*, 9:116-119 (Russian with English summary).
Stepanov, G.S. (1977). "Breeding value of different types of intervarietal hemp hybrids." *Selektsya i semen.* In *Referativny Journal*, 11.55.359 (Russian).
Streletsky, A. V. (1970). "Hemp breeding and seed production in Poland." In *Len i Konoplya* (Flax and Hemp), 1970, 10:37 (Russian).
Tran Van, Lai (1985). "Effect of inbreeding on some major characteristics of hemp." *Acta Agron Acad Sci Hung*, 34:77-84.
Vedenski, V., (1929). "On the methodic and actual problems of fiber hemp breeding." *Bull Appl Bot Gen Plant Breed*, Leningrad.
Virovets, V. (1997). Personal communication.
Virovets, V. G., Gorshkova, L. M., and Sitnik, V. P. (1980). "New monoecious hemp varieties." *Len i Konoplya* (Flax and Hemp), 6:28-29. (Russian with English summary).
Virovets, V. G., Lepskaya, L. A. (1983). "Varietal resistance in hemp to European corn borer (*Ostrinia nubilalis*, Hb.)." In *Referativny Journal* (1985), 12. 65.390.
Virovets, V. G., Stserbanj, I. I., and Sitnik, T. I. (1976). "On seed productivity of the monoecious hemp hybrids." *Biol. vozd. i pervich. obrobt. kon. i kenafa*, Gluchov, 39:21-26.

Chapter 9

Advances in Biotechnological Approaches for Hemp Breeding and Industry

Giuseppe Mandolino
Paolo Ranalli

The term plant biotechnology covers a collection of techniques for the manipulation of plant cells, tissues, or molecules. As far as plant tissue and organ culture are concerned, the related techniques, involving essentially aseptic culture of plant parts, have been developing since the 1940s, while the techniques for the manipulation of molecules such as DNA or RNA only became available in the last fifteen years, and have been extensively applied to plant research during the 1980s.

There have been various fields of application of plant tissue culture, but probably the most successful ones are the breeding and production of useful secondary metabolites. In both cases, it would be expected that the applications to hemp would lead to remarkable results. These two main fields of plant biotechnology and the results that can be expected for hemp cultivation will be briefly reviewed in the first two paragraphs. The subsequent development of molecular techniques and the impact that can be foreseen will be discussed in the third part of this chapter. Finally, the possible future perspectives of genetic engineering in hemp—a technique combining the characters of plant tissue culture and of molecular manipulation—will be highlighted.

TISSUE CULTURE AND BREEDING

According to the definition of Steward (1983), the "art of growing aseptically and heterotrophically isolated plant portions as separated explants on chemically defined media has become identified with plant tissue culture." The term "art" used by Steward appears justified by the fact that the basic mechanisms underlying the different responses of the explant to a defined medium are largely unknown, and tissue culture protocols are only in part based on the knowledge of the cause-effect relationship. However, despite the very large fluctuations of responses with the age of the donor plant, the type of explant used, the species considered and, within a species, the genotype itself, plant tissue culturists have over the years established a few guidelines that can be considered when a tissue culture research program is started on a species not considered before.

The clonal propagation of plants, when performed in aseptic conditions on synthetic media, is termed micropropagation. This technique has proved to be very useful and economically convenient in several plant species, as a means for obtaining a large number of genetically identical plants in short times. The advantages of micropropagation on conventional practices of asexual propagation are manifold: (1) a tiny amount of initial tissue is able to generate a high number of plants; (2) cultured shoot tips or buds are more suitable for handling, expedition, exchange, and conservation; (3) aseptic culture ensures the plant material is sanitary; (4) the multiplication of the *in vitro* stock can be performed at any time of the year, allowing a greater flexibility of the cultural cycle; and (5) it has often been reported that *in vitro* propagation of meristem culture, combined with thermal treatments gets the initial material rid of some virus diseases (the example of certified seed potatoes obtained by micropropagation is the most known one; Stace-Smith and Mellor, 1968; Pennazio and Redolfi, 1974; Mellor and Stace-Smith, 1977).

To the best of our knowledge, there are no experimental demonstrations that hemp viral diseases can be eradicated by shoot culture techniques, as is known for a wide number of other species (Hu and Wang, 1983). For *Cannabis* species the real problem appears to be the feasibility of micropropagation itself, while practices of clonal propagation of hemp are well known (Bizarri and Ranalli, 1995). In a species like hemp, with such a high degree of variability, the use of propagation techniques, allowing the production of a large number of genetically homogeneous plants, is extremely useful. Among the crucial steps of breeding hemp cultivars are the control and analysis of the content of Δ^9-tetrahydrocannabinol (Δ^9-THC), or of fiber content; these analyses must be performed on defined tissues and at defined moments of the plant development. The availability of cloning techniques is therefore useful for the conservation and multiplication of plants to be used subsequently for crossing and seed production of materials endowed with low levels of Δ^9-THC or high fiber content or quality. The availability of clonal propagation protocols is also of crucial importance when mutagenesis experiments have been carried out, in order to identify hemp mutants with valuable characters; in most cases, in fact, the fixation of the (probably) recessive mutant character requires the crossing of two mutant plants, that can be obtained by asexual propagation.

One possible approach (Bizarri and Ranalli, 1995) is to detach stem segments, 10 to 15 cm long from the mother plant. These cuttings are then transferred in Erlenmeyer flasks containing 75 ml of MS medium (Murashige and Skoog, 1962). In these experiments, the stem segments were placed in MS medium containing different growth regulators (see Table 9.1). The table also reports the success of radication induced in this way on the cuttings (see Figure 9.1). The effect of the indolbutirric acid (IBA), known in other plant systems to induce the rooting of the cultured shoots, is to induce root formation in a high percent of cases; this effect is not mimicked by naphtalenic acid (NAA), which appears to antagonize the root induction generated by IBA. The rooted segments can easily be transferred to soil or field with a high percent of survival; results on the same technique have been published by Richez-Dumanois et al. (1986).

The previously described procedure, although very useful when only a few plants are needed, cannot be successfully applied when hundreds of genetically identical plants are needed in a short time. In these cases, *in vitro* techniques must be applied, involving essentially shoot or bud culture, or shoot regeneration from cultured explants or callii.

Table 9.1. Liquid media used to test *in vivo* rooting of hemp stems.

Medium Number	Composition	Rooting %
1	MS*	0%
2	MS + IBA 10^{-4} M	70%
3	MS + NAA 10^{-4} M	19%
4	MS + IBA 10^{-4} M + NAA 10^{-4} M	44%

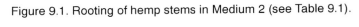

Source: T. Murashige and F. Skoog. (1962)."A revised medium for rapid growth and bioassays with tobacco tissue cultures." *Physiologia Plantarum*, 15:473-497.

Figure 9.1. Rooting of hemp stems in Medium 2 (see Table 9.1).

Very little has been published on these topics; Richez-Dumanois et al. (1986) report that, on two fiber cultivars (F77 and F56), the choice of the primary explant is crucial; these authors used donor plants at least 1 m long. After sterilization of the buds, they placed them in solid MS medium supplemented with either glucose or sucrose, and with activated charcoal 2 g/l. In these conditions, under a photoperiod of 16:8 hours, the meristems develop enough to be subcultured every three weeks, obtaining an increasingly higher number of nodal cuttings. The growth of the hemp cuttings does require the presence of growth regulators. Richez-Dumanois et al. (1986) report that in general, an auxin/cytokinin ratio below 1 induces the development of the axillary buds; in particular, benzyladenine (BA) at 5×10^{-7} M (about 0.45 mg/l) and IBA at 10^{-7} M 20 (μg/l) are used for both the culture of the initial buds and for subsequent proliferation. NAA in place of IBA is reported to favor the formation of a basal callus. The elongation of the nodal cuttings seems to require active charcoal 2 g/l and IBA 2 mg/l; rooting was described to be extremely difficult, with the use of kinetin 10^{-7} M and IBA 10^{-5} M giving a rooting reported to be "possible, but aleatory." To some extent, these results might be due to the choice of the basal culture medium, MS. In our experience, better results are obtainable with B5 medium (Gamborg, 1966).

When the primary explant, consisting of a 0.5 cm stem segment carrying two axillary buds, is placed in MS medium, poor or no growth is observed, while in B5 medium about 70 percent of the explants start growing; the growth is however not uniform for all cuttings, and no rooting occurs if the basal medium only is employed. The hemp plantlets can be propagated about every three or four weeks, but unfortunately, as the propagation cycles are repeated over and over, the growth becomes slower, and an increasing number of nodal cuttings cease to develop, turn yellow, and die. If the B5 medium is supplemented with gibberellins in the range 2 to 5 mg/l, the explants do not develop at all, as already observed by Richez-Dumanois et al. (1986), while the addition of IBA 20 mg/l efficiently stimulates the rhizogenesis, though in the subsequent subcultures the yellowing effect already observed is markedly enhanced (Moschella, unpublished).

Another approach very useful in obtaining a clonal population of plants through tissue culture is the regeneration of shoots from plant explants or from callus deriving from cultured plant segments. Callus appears a promising material to deal with, as it was easily established and propagated, and in a limited time a high number of petri dishes containing growing callii can be obtained. Experiments on the conditions best suited for establishing a callus culture in hemp (Zottini et al., 1996; Moschella, Mandolino, and Ranalli, 1996) showed, as expected, a genotype and type of explant dependance of the callus abundance and type obtained. All cultivars or accessions tested were indeed able to form callus, but the feasibility of callus subculturing and propagating was different, ranging from the appearance of small amounts of brown callus in some genotypes, to an abundant production of white, friable callus in others. The media used to induce callus was MS basal medium, supplemented with B5 vitamins, sucrose 3 percent, and different types of growth regulators, depending on the explant source (see Table 9.2). This suggests hemp has a good capacity to generate undifferentiated tissue, a capacity that has in the last twenty years been exploited in order to obtain callus-deriving suspension cultures, useful for the study of *Cannabis* secondary metabolism.

Unfortunately, we found a very poor ability of the proliferating callus, when plated on media with an auxin/cytokinin ratio below 1 (reported in several species to promote shoot regeneration), to yield shoot formation. This last phenomenon is at best occasional at the moment, though among twelve cultivars and accessions tested, one (from Gatersleben germplasm collection, accession CAN 19/86, "Sud Italian") occasionally gave rise to strongly organogenetic callus, if 2,4-D was present in the medium. The regeneration events we observed (Moschella, Mandolino, and Ranalli, 1996) start with the formation of green organized structures in the callus mass (see Figure 9.2a). The growth regulators used to induce this phenomenon were 2,4-D in the range 3 to 10 mg/l, and BA 0.01 to 0.1 mg/l. For all combinations in this range, shoot formation was observed, except in leaf-deriving callus (Table 9.2; Figures 9.2b-f). The high concentration of 2,4-D used was effective in stimulating shoot formation within ten days from the transfer of the callus; after this period, if no regeneration was evident, the callus had to be transferred to a low 2,4-D medium, otherwise browning and degeneration occurred. It should be noted, however, that the data reported in Table 9.2 are largely preliminary, to be taken as guidelines for further experiments.

Table 9.2. Callus induction and shoot regeneration ability as a function of the type of explant.

Explant source	Callus formation	Shoot regeneration
Leaf	+ + +	−
Hypocotyl	+ + +	+ + +
Cotyledon	+	+ +
Root	+	+

Note: −: no event observed; +: occasional event observed; + +: different events observed; + + +: abundant callus or shoot formation.

Figure 9.2. *In vitro* behavior observed with cultured hemp explants: (a) Callus development; (b) initial shoot formation; (c-d) shoot development completed; (e) subcultured hemp shoot; (f) developed shoot prior to transfer for rooting.

b

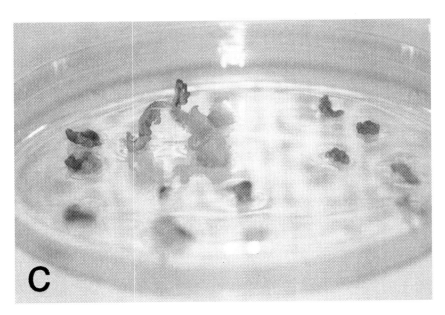

c

Advances in Biotechnological Approaches

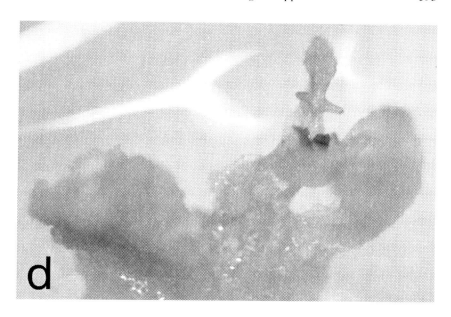

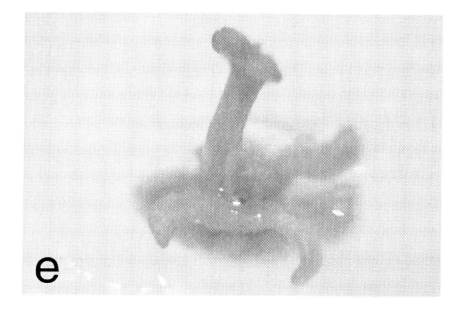

To our knowledge, the above described events constitute the first report of shoot regeneration in *Cannabis* species. We are still far from erasing hemp from the list of those plant species considered to be "recalcitrant" to *in vitro* techniques, and much work still has to be done, but the research should now focus on the conditions rendering the *in vitro* propagation and the shoot regeneration less aleatory and genotype-dependent.

CELL CULTURE AND SECONDARY METABOLISM

In nature, the phenomenon of plant cell growth and division is a strictly regulated one, depending upon a number of stimuli, both endogenous and exogenous. The technique of plant cell culture allows the cells to divide and grow continuously, and has in the last years been employed to obtain commercial and medicinal compounds, and to study cellular metabolism as well. In most cases, the compounds of economic value are secondary metabolites, like pigments, alkaloids, steroids, etc. The problem of cell cultures for commercial purpose is that, in most cases, secondary metabolites are produced by the cultured cells in only small amounts, often to a much lower extent than in intact plant tissues. Despite this drawback, the perspective of creating large-scale cell cultures to be used as bioreactors stimulated the search for high-yield cell lines (Yamada, 1984); this approach has already been successful for several compounds, and patents have been issued for the production by plant cell cultures of several metabolites (Staba, 1977).

It has been pointed out that the commercial availability of crude plant material from, for example, politically unstable third world countries, might, however, be difficult, and in these cases the use of large-scale cell cultures is justified (Dodds and Roberts, 1982). This is not actually the case in hemp, but nevertheless in the last twenty years the potential of *Cannabis* cell culture has been carefully examined.

As stated by Pate (1994), hemp is a virtual factory for the production of secondary metabolites, producing alkanes, nitrogenous compounds, flavonoids, terpenes, and other compounds. The taxonomy of hemp and the differentiation of the types is largely based on the production of secondary substances, among which the most typical are the cannabinoids (see Chapter 3).

The first step in the evaluation of the feasibility of secondary product purification from plant cell cultures is obviously the observation that the cells do synthesize the compound of interest. In 1972, it was reported that suspension cultures of *Cannabis sativa*, established from root-deriving callus, were not able to produce any cannabinoids (Veliky and Genest, 1972). This is not surprising, as it is known that the roots do not contain the epidermal glands, where the biosynthesis of cannabinoids occur. This report could be considered simply as a hint that the block of cannabinoid production, acting in the root cells, is maintained in cultured cell systems, deriving from root tissue. In subsequent experiments, callus was induced from different organs, such as maturing bracts and anther-calyx complexes after anthesis, from both fiber and drug hemp types (Hemphill, Turner, and Mahlberg, 1978). Callus was initiated in a modified Miller's medium (Miller, 1968), supplemented with either IAA (indolacetic acid) 1 mg/l and kinetin in the range 1 to 1.5 mg/l, or with IAA 0.25 mg/l, NAA 0.25 mg/l, 2,4-D 0.2 mg/l, and kinetin 1 mg/l. The formation of roots from the callus was observed, but it was found that 2,4-D presence inhibited this phenomenon in freshly formed callus; callii subcultured in maintenance medium (Miller's medium supplemented with IAA 1 mg/l, NAA 0.1 mg/l, 2,4-D 0.2 mg/l, and kinetin 2 mg/l) did not form roots. Interestingly, these authors found that casein hydrolisate (CH) enhanced the callus growth, but, except for the calix callus of a fiber hemp type, they were unable to mimick this effect by substituting CH with defined aminoacid mixtures. The supplemented aminoacid seemed however to play a role in the biosynthesis of secondary metabolites similar to cannabinoid standards, at least in bract-deriving callus. In this case, an aminoacid mixture composed of glycine, glutammate, methionine, phenylalanine and aspartate (all in their L-forms) promoted the synthesis *in vitro* of several compounds chemically related to cannabinoids (as estimated by gas-liquid and thin-layer chromatography), while the use of Gamborg's aminoacid mixture (Gamborg, 1966) did not yield any compound chemically comparable to cannabinoid standards. However, none of the major cannabinoids detected in the mature plant were found to be present in the callus cultures.

The induction of the biosynthesis of secondary compounds in other plant species, such as *Nicotiana tabacum* for nicotine production, appears to be strictly related to organ differentiation (Staba, 1977; Peters et al., 1974); if this holds true for hemp too, the necessity arises of establishing long-term cultures of organogenetic callus or cell suspensions, in order to exploit the possibilities of hemp cells as bioreactors. Occasional rooting or shoot formation events do not take place in a sufficiently large number of callii to verify this hypothesis; further experimentation seems necessary in order to isolate organogenetic cell lines, similar to what is reported in other species.

The first report of successful synthesis of cannabinoids by an *in vitro* cellular system in hemp came fifteen years ago (Heitrich and Binder, 1982); later on, it was shown by other authors that cannabidiol could be converted to cannabielsoin by a hemp cell suspension (Hartsel, Loh, and Robertson, 1983). Braemer and Paris (1987) showed that a cell suspension culture of *Cannabis sativa*, deriving from leaf callus, and kept in culture for eight years, was able, within six days from the addition of cannabidiol, to completely convert it to cannabielsoin and cannabielsoin-1,8 cineol derivative, although to low yields. The same kinetics of disappearance was reported when Δ^9-THC was added as substrate. Interestingly, the cell culture was also in this case established in B5 medium, supplemented with kinetin 0.5 mg/l and 2,4-D 1 mg/l, while the substrate used was glucose. In fact, it was known that hemp suspension cultures could grow to a good rate in the presence of a number of carbon sources different from sucrose, i.e., glucose, fructose, galactose, and glycerol (Jones and Veliky, 1980). This finding is important in view of the use of hemp cells in industry, as in this case a cheaper and more easily available substrate than sucrose is needed.

Another report of successful biotransformation of cannabidiol by *Cannabis* suspension cultures was from Loh, Hartzel, and Robertson (1983); these cultures, inoculated with cannabidiol, produced two major cannabinoids, identified by gas chromatography and mass spectral analysis as diastereoisomers of cannabielsoin, while other unidentified cannabinoids were produced by the culture when olivetol was used as substrate. The capacity to metabolize the added substrate seems related to the maturity of the cell culture, as biotransformation starts as soon as the culture reaches the stationary phase, at approximately 13 mg of cellular protein per ml of culture, after about fourteen days from each subculture.

These examples show that it should be possible in the future to obtain cannabinoids or other valuable related products by using *in vitro* cultured hemp cells; this system also appears to be of potential utility for the study of hemp metabolism, especially if the *in vitro* techniques are implemented with the more recently developed tools from molecular biology.

MOLECULAR MARKERS FOR HEMP BREEDING

During the last fifteen years, one of the fields of plant biology that has been more widely explored, registering impressive advancements, is the molecular mapping of plant genomes. Only recently have the molecular tools developed enough to allow the exploration —and exploitation—of the enormous amount of genetic polymorphism present in plant natural populations, known since the 1960s. This latter finding was one of the major cultural breakthroughs in the field of genetics, but it was only with the advent of recombinant DNA technology that this important concept found its full application.

The degree of polymorphism of a plant species is the basis of the possibility to perform linkage analysis. In plants, differences at the DNA level vary greatly in the different species: a study of Shattuck-Eidens et al. (1990) showed a variation of 1.2 percent in maize, and of 0.02 percent in melon (calculated on the basis of DNA sequence variation data). The first question that the application of DNA technology in *Cannabis* should address is therefore: how great is the level of DNA polymorphism within the species? This problem is of paramount importance in programming, for example, a linkage analysis of *Cannabis sativa*. In general, species that are self-pollinating are expected to have a lower level of polymorphism than the outcrossing species like hemp, and indeed there are experimental data in this direction (Helentjaris et al., 1985). The approach followed in these experiments was to sequence different wild-type alleles at the same locus, and directly verify the number of variant basepairs.

The observation that some phenotypic characters are closely linked to an agronomically important trait, such as disease resistance or productive characteristics, came early in this century. The extensive use of marker characters, however, requires the construction of genetic maps, in which such phenotypic markers are linearly ordered on linkage groups. The drawback of this approach relies on the limited number of available markers in most species; a phenotypical trait, in order to be used, must be visible or easily scored, and therefore depends upon the expression of a particular allele at a given locus. A linkage group, representing a chromosome, cannot be saturated with phenotypical markers, and consequently the position of an agronomically or economically important locus cannot in most cases be determined with precision. Moreover, the higher the distance between any marker locus and the locus of interest, the higher the probability will be of breakage of the association, due to recombination. In 1980, it was realized that a genetic map could be constructed by using segments of chromosomal DNA as markers, and that the segregation of this type of markers could be followed in the progeny, and hence function as Mendelian alleles, codominant and simply inherited (Botstein et al., 1980). These new types of markers were called RFLP (Restriction Fragment Length Polymorphism), because they could be generated by digestion of two parental DNAs with restriction endonucleases. These enzymes, today commercially available, cut the DNA in specific sites (usually in correspondence of six-basepairs long target sequences), that can be easily separated by agarose gel electrophoresis. The possible polymorphism existing between the two parental DNAs can, with some probability, involve the target sequences of a given enzyme; this latter will therefore cut the DNAs in different ways, and different sized bands will migrate in different positions in the gel. Gel electrophoresis is usually followed by a capillary transfer of the DNA fragments on nylon or nitrocellulose membranes (Southern blotting; Southern, 1975); the membranes can then be hybridized with short DNA sequences, homologous or also heterologous, radioactively labeled (acting as probes), in order to be able to identify the corresponding fragments on the membrane, after autoradiography. This technique, despite its complexity and costliness, and need for rather sophisticated laboratory equipment and expertise, became widely used to detect genetic polymorphisms, as its great power relied on the fact that differences in the DNA could also be detected in genome regions not expressed. The consequence of this advantage is that the number of RFLP markers is enormous, as the combinations of probes and restriction enzymes that can be used is theoretically unlimited. The introduction of RFLP technique in the plant genetics was a major breakthrough, and opened the way to the construction of highly saturated maps, including hundreds of molecular markers distributed in linkage groups, associable to phenotypic, productive, or resistance traits. RFLP maps of many important crop species are now available, including molecular markers linked not only to Mendelian characters, but also to quantitative trait loci (Tanskley et al., 1989; Knapp, 1994).

In plant research, another approach, based on the recently introduced PCR (Polymerase Chain Reaction) technique, has became prevalent on RFLP. Among the PCR based techniques, the RAPD (Random Amplified Polymorphic DNA; Williams et al., 1990) strategy of fingerprinting and mapping plant genomes has been the most widely applied, probably because of its extreme simplicity, the fact that it requires minimal laboratory equipment, and does not involve the handling of radioactive material. In this technique, short (usually ten nucleotides long) single strand oligonucleotides are used; these oligonucleotides can act as primers in starting the genomic DNA amplification process mediated by a particular enzyme, the *Taq* polymerase. In the plant DNA, the primers frequently recognize identical sequences opposed to each other, and located in the DNA stretch at a distance allowing the repeated synthesis of the intervening sequence. The result in this case will be the exponential amplification of discrete parts of the genome, that will appear as intense bands in agarose gel electrophoresis; in this case, no blotting or hybridization procedure is needed. Two genotypes can be polymorphic in the distance occurring between two of the attachment sites of the primers, and will therefore exhibit a different set of amplified DNA products, scorable as visible bands in the gel. The markers can again be scored as Mendelian factors, but in this case will usually be dominant, and therefore expected to segregate in the progeny with a 3:1 ratio.

Despite the fact that RAPD technique has seldom been applied in animal research, it has proved to be of great potential for mapping plant genomes and specific characters. Another advantage of RAPD, and in general of PCR-based techniques, on RFLP, is that only a tiny amount of genomic DNA is needed for each assay. While 10 to 15 μg of DNA are necessary for an RFLP analysis, only 10 to 50 ng are commonly employed for each amplification reaction using RAPD.

The first step to carry out when an RFLP or RAPD analysis program is being started on a new species, is the adaptation of the protocols of genomic DNA extraction to that species. This step did not present any particular problems in *Cannabis sativa*, either of fiber or drug type. Despite the high content of secondary products, typical of this genus, the standard protocols found in the literature applied fairly well, and can be considered suitable, at least for restriction analysis with most endonucleases, amplification by *Taq* polymerase, and cloning in plasmid vectors. For all these applications, a further purification of DNA in CsCl gradients was not found to be necessary. The standard protocols adopted in the few papers published on molecular markers in *Cannabis* (Sakamoto et al., 1995; Faeti, Mandolino, and Ranalli, 1996) essentially consist of the following steps (Doyle and Doyle, 1990; Faeti, Mandolino, and Ranalli, 1996):

1. grinding of the plant tissue under liquid nitrogen;
2. resuspension of the powder in an extraction medium, consisting of a buffer at pH 8.0 and a detergent such as SDS (sodium dodecylsulphate) or CTAB (cetrimide), able to eliminate glyco-

proteins or other secondary products; at this stage, proteinase K can be added to destroy enzymatic activities that can damage the DNA, but in our experience this digestion step can be omitted;
3. phenol and phenol-chloroform extractions, to obtain an acqueous, clean phase with the DNA in solution; and
4. isopropanol and/or ethanol precipitation.

These protocols are already a simplification of the earliest methods, but if a large number of individual plants have to be assayed, as usually happens when a segregating progeny is being tested, a further scaling-down, and possibly automation, of the procedure may be advisable. If the experiments involve PCR-based techniques, microextraction methods are particularly useful; PCR in fact allows the analysis of hundreds of DNA samples per day, after the DNA has been purified. In this case, it is the preparation of intact genomic DNA which is the time-limiting factor. Several reports have been published dealing with the development and large-scale application of microextraction methods for plant DNA (Deragon and Landry, 1992; Cheung, Hubert, and Landry, 1993; Klimyuk et al., 1993). In most cases, these methods have been successfully applied, require only very small amounts of tissue, avoid the dangerous and expensive use of liquid nitrogen, and yield restrictable and amplifiable DNA for a large majority of samples; however, there are no reports of their applicability in hemp.

DNA technologies have only recently been applied to *Cannabis sativa* and in related dioecious species of the Cannabaceae family. In hop *(Humulus lupulus)*, the level of genetic variation as estimated by RAPD analysis was found to be low; out of sixty decamer primers used, only eight produced polymorphic RAPD bands (Pillay and Kenny, 1996). In this study, it was shown that nineteen primers failed to produce any amplified DNA at all, and only for three primers was it possible to follow the segregation of the markers in the progeny. This low level of polymorphism was, however, higher than estimated by RFLP, reported to show no polymorphism at all. The reason for this homogeneity could be that hop varieties were selections endowed with phenotypically superior qualities, propagated since the early hop breeding practices. Breeding hemp is a different practice, and the polymorphism levels detected even in genera belonging to the same family may be very different.

The knowledge of the extent of genetic variation within a species is important for several reasons. First, it provides a basis for the development of new genotypes; the existence of a strong variability among genotypes cariologically homogeneous—and hence interfertile—is a potential advantage for the breeder, enabling him to select parental lines from a wider genetic pool. In hemp, breeding for increased stem yield and quality, and for decreased Δ^9-THC, are among the main targets to be pursued. In this framework, the constitution and the evaluation of germplasm collections has assumed increasing importance (de Meijer and van Soest, 1992; see Chapter 7). The use of molecular markers is becoming widespread in the characterization and screening of these collections. In a recent paper, Faeti, Mandolino, and Ranalli (1996) examined the genetic variability found in cultivars and accessions from different sources. The technique used in this case was the RAPD analysis, more suitable to screen a wide number of samples in short times. In this study, fifty-four plants belonging to cultivars (Carmagnola, Fibranova, Bialobrzeskie, Beniko, Kompolti Sárgaszárú) or accessions (four accessions from Hungary, one from Korea, one from the Czech Republic, one from Rumania, one from Italy) were examined. Ten different decamer primers were used, all commercially available. The results were impressive. The average number of bands produced by PCR ranged from five to thirteen, and no sample with any primer failed to show detectable bands; the RAPD profiles obtained were in the large majority of cases highly reproducible, even when not of full intensity, and easily scorable (see Figure 9.3). The level of polymorphism found in these experiments was exceptionally high, in contrast to what is described for hop: out of 205 total markers scored in the fifty-four plants, only four (i.e., about 2 percent) were not polymorphic across all the individual plants of the cultivars and accessions tested.

These results indicate that in hemp germplasm an above-average polymorphism is present at the DNA level, or at least that the polymorphism present is easily detectable by RAPD analysis. Even within two groups of plants belonging to the Italian cultivars Carmagnola and Fibranova, it was reported that only 20 percent of the DNA bands were monomorphic, suggesting an only slightly lower level of polymorphism within germplasm subjected in recent times to selection; there is however a high degree of variability. Large genetic differences were

Figure 9.3. RAPD pattern observed on ten Carmagnola (lanes 1-5 male, lanes 6-10 female) and ten Fibranova (lanes 11-15 male, lanes 16-20 female) DNAs. The arrow indicates the male specific band; the asterisk shows one of the deviant phenotypes (see also Table 9.3).

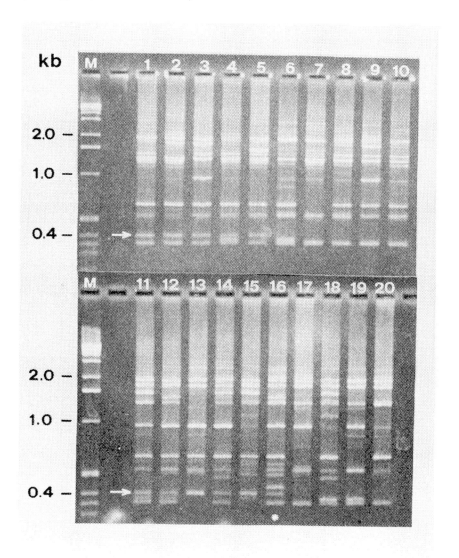

evident between accessions; the RAPD approach allowed a tentative classification into groups of the germplasm studied, on the basis of agenetic similarity index (Dice, 1945). It was observed that the four groups identified are coincident to a good extent with the geographical origin of the material studied; one gene pool appeared to group the Italian cultivars, another cluster is formed by Hungarian material, and a third group by plants of Asian origin. A fourth group appears less defined, formed of miscellaneous material from Italy, Poland, the Czech Republic, and Rumania; interestingly, the monoecious plants tested formed a small subcluster within this group.

In these experiments it was demonstrated that, given the stability and reproducibility of the RAPD profiles, and the high degree of polymorphism found in *Cannabis*, the molecular techniques can be profitably employed also in this species to check the diversity of the material for breeding purposes; in addition, such a high degree of DNA polymorphism suggests that the molecular mapping of the hemp genome could be performed, using RFLP or RAPD technique. In view of this, an evaluation of RAPD profiles of drug type hemp was also made (Mandolino, unpublished data), showing a markedly lower degree of polymorphism compared to the fiber types examined so far, but probably still sufficient to be used for mapping purposes.

Hemp belongs to the 10 percent of Angiosperm developing unisexual flowers; in nature, it is dioecious, the male and female flowers being carried by different plants. Consequently, *Cannabis* species is an obligate outbred. Unlike other dioecious species in which during the early stages of flower development, primordia are present in both male and female structures, followed by arrest or abortion of one of the two types, in hemp this sort of early hermaphrodite stage seems to be absent (Mohan Ram and Nath, 1964). The genetics of sex determination is very different in different plants; in some species one single locus is responsible for the sex differentiation (cucurbit), in others several unlinked loci are needed (cucumber), and in others a whole linkage group, acting as a sex chromosome, is necessary. Male plants are in most cases heterogametic, but there are exceptions to this rule (see Lebel-Hardenack and Grant, 1997, for a review).

In *Cannabis sativa,* two heteromorphic sex chromosomes have been described in the past (Hirata, 1924); according to later studies of Yamada (1943), hemp female plants have two X chromosomes and male plants have one X and one Y chromosome, this latter being larger than X. Autosomic genes are not involved in sex determination, according to Warmke and Davidson (1944); this scheme would be a typical active-Y system, but it should be pointed out that in subsequent years the presence of heterochromosomes in hemp, and the type of sex determination, was questioned (Moham Ram and Sett, 1985). Regardless of the presence of heterochromosomes, however, other authors proposed the existence of masculinizing alleles, named Xm, also on the X chromosome (Durand and Durand, 1990). To further complicate the scheme of sexual determination in hemp, it is known that, upon chemical treatments, some female genotypes can be induced to form viable pollen, while other genotypes appear to lack this possibility (de Meijer, personal communication); environmental factors and metal ion treatment was also reported to allow sex reversal in this plant.

In dioecious plants, the early determination of the sexual phenotype of a plant and the possibility of screening progenies for sex segregation may have particular importance for the breeder. It is not surprising, therefore, that the most active research efforts in dioecious species focus on the search for molecular markers closely linked to the sex phenotype. Particularly in hemp, the determination of the gender might be useful in breeding programs. The most known Italian and Hungarian cultivars are dioecious, but also where the breeding focused especially on the development of monoecious varieties, the possibility of an early identification of the sex would be of great use in maintaining the monoecious character during the crosses. According to Bócsa (1994), monoeciousness has the disadvantage of a slower rate of genetic improvement, because the stem yield will be 10 to 20 percent lower than in dioecious hemp— due to inbreeding—and because the Bredemann method cannot be applied. This latter method involves the early scoring of the fiber content in the male plants, followed by pollination of the female plants only from those males with a superior fiber content or quality. Whatever the breeding strategy chosen, dioecious or monoecious hemp, the molecular markers presently already provide tools in the direction of an early prediction of the gender. The first published report of such a marker was from Sakamoto et al. (1995); in this paper, two male-specific bands were described, generated with the RAPD technique by two different decamer primers (with the sequences: 5'-ATCCGCGTTC-3' and 5'-ACGGCATATG-3'). The bands were 500 and 700 bp long, and were present in the five male plants and absent in the five female plants tested (CBDA strain). This finding was followed by other results from our laboratory, working on DNA variability in a wide hemp germplasm, as described above (Mandolino et al., 1995; Mandolino et al., in press). In this last case, it is interesting to note that the finding arose from a previous study on genetic diversity in *Cannabis sativa*. It was found that, within one of the clusters obtained by UPGMA (Unweighted Pair Group Method with Arithmetical Analysis), the ten female plants of the two Italian cultivars Carmagnola and Fibranova, tended to form a separate subcluster from the ten male plants of the same two cultivars. Apparently therefore, with the markers scored with ten random primers used in that study (Faeti, Mandolino, and Ranalli, 1996), the gender similarity was superior to the variety similarity. This suggested the possible presence of different gender-specific markers. The hypothesis was confirmed by the subsequent analysis, carried out on a larger number of individuals, with a known sexual phenotype. Different primers generated male-specific bands; among these, the most reliable and easily scorable was a 400 bp band generated by the decamer primer with the sequence 5'-GTGACGTAGG-3' (see Figure 9.3). A wide survey of cultivars and accessions revealed that, out of fifty-nine male plants tested (belonging to twelve different genotypes), the 400 bp marker was always present, while in the eighty-eight female plants examined, the marker was present in only three cases (two Carmagnola and one Fibranova plant; see Table 9.3). Drug-type hemp lacked the 400 bp marker, as well as plants of the monoecious cultivars Beniko and Bialobrzeskie (supplied by Professor R. Kozlowski, Institute of Natural Fibers, Poland); however, the progeny of open-pollinated monoecious plants showed the expected sex segregation, and this segregation was precisely linked with the appearance of the 400 bp marker in the male plants (Mandolino, unpublished). The male-specific markers obviously have the characteristic of being codominant, and not dominant as most of the RAPDmarkers. Sakamoto et al. (1995) and Mandolino et al. (in press) isolated, cloned, and sequenced the DNA fragments associated with the male phenotype; the two sequences are different, but share the $G+C$ content (about 39 percent) and the fact that they are not expressed sequences, as in both cases no open-reading frames were present in all frames and in both orientations.

Table 9.3. Presence of the 400 bp marker in the hemp plants examined.

Cultivar or accession	Male plants	400 bp marker	Female plants	400 bp marker
Carmagnola	26	26	36	2
Fibranova	10	10	22	1
Uniko	5	5	5	—
Kompolti Hybrid TC	5	5	5	—
Kompolti Sárgaszárú	5	5	5	—
Accession 1	—	—	4	—
Accession 2	3	3	3	—
Accession 3	2	2	2	—
Accession CAN21/86	1	1	2	—
Accession CAN16/78	1	1	1	—
Accession CAN17/81	1	1	1	—
Accession CAN19/86	—	—	2	—
TOTAL	**59**	**59**	**88**	**3**

The occurrence of different male-specific RAPD markers seems a characteristic of hemp; in fact, it is in sharp contrast with reports from other dioecious species, where either no sex-linked markers were found, as in hop (Pillay and Kenny, 1996), or a very large number of RAPD primers or RFLP probes had to be screened before finding a loose association with the gender, as in *Pistacia vera* (Hormaza, Dollo, and Polito, 1994) or in *Asparagus officinalis* (Restivo et al., 1995). Apparently, a much wider region of the genome is involved in sex determination in hemp; alternatively, the commercially available decamer primers used in RAPD analysis have a strong degree of affinity for male-associated DNA regions.

The described markers are immediately utilizable by the hemp breeder; a future field of study will be the use of these markers not only for practical purposes, but also for the investigation of the molecular mechanisms of sexual differentiation in *Cannabis sativa*.

FUTURE PERSPECTIVES

In this short review, we tried to outline the state of the art of hemp biotechnology. It is apparent that there are fields in which a gap exists between this species and others; reliable and reproducible *in vitro* techniques are still a long way off, and unfortunately, because these techniques form the basis for genetic engineering, the perspectives for genetic transformation are not immediate. One priority is therefore, in our opinion, to pursue these types of studies; as far as we know, no results have yet been obtained in this important field of biotechnology. Hemp must still be considered a recalcitrant species, and major efforts are needed in this direction. The availability of transformation techniques would allow the rapid development of new cultivars endowed with, for instance, visible markers, useful for the discrimination of fiber and drug types, or with disease resistant traits.

The availability of cell cultures producing valuable secondary metabolites, combined with the most recent development of molecular tools, should in the near future allow the identification of genes controlling the synthesis of these products, for instance Δ^9-THC. The study of the structure of these genes, and of the relative controlling sequences (enhancers, promoters) might lead to an alteration in the constitutive level of secondary products synthesis, with interesting applications of the cell cultures in the industry.

Finally, the availability of molecular markers for specific traits, gathered together in a saturated linkage map would have a remarkable impact on hemp breeding. The perspectives seem to be good, given the high level of polymorphism at the DNA level already observed; the close association found between male phenotype and specific molecular markers, even in the absence of a molecular map, can already become a rapid diagnostic tool in breeding. However, saturated molecular maps will be needed in order to find markers associated with other important characters, probably quantitative in nature, such as Δ^9-THC and fiber content; in these cases the degree of heritability will influence the selection method to be chosen. Mapping these QTL is certainly the most important task for the future; joint research programs, with the participation of the seed or textile industry, will probably play a major role, as already demonstrated in the recent past for other crop species.

REFERENCES

Bizarri, M. and Ranalli, P. (1995). "Propagazione della canapa da fibra (*Cannabis sativa* L.) per talea." *L'Informatore Agrario*, 26:57-59.
Bócsa, I. (1994). "Interview." *Journal of the Internationl Hemp Association*, 1(2): 61-63.
Botstein, D., White, R. L., Skolnick, M., and Davis, R. W. (1980). "Construction of a genetic linkage map in man using restriction fragment length polymorphisms." *American Journal of Human Genetics*, 32:314-331.
Braemer, R. and Paris, M. (1987). "Biotransformation of cannabinoids by a cell suspension culture of *Cannabis sativa* L." *Plant Cell Reports*, 6:150-152.
Cheung, W. Y., Hubert, N., and Landry, B.S. (1993). "A simple and rapid DNA microextraction method for plant, animal, and insect suitable for RAPD and other PCR analyses." *PCR Methods and Applications*, 3:69-70.
Deragon, J. M. and Landry, B. S. (1992). "RAPD and other PCR-based analyses of plant genomes using DNA extracted from small leaf disks." *PCR Methods and Applications*, 1:175-180.
Dice, J. R. (1945). "Measures of the amount of ecologic association between species." *Ecology*, 26:297-302.
Dodds, J. H. and Roberts, L. W. (1982). "Production of secondary metabolites by cell cultures." In *Experiments in plant tissue culture*, pp. 141-148, New York: Cambridge University Press.
Doyle, J. J. and Doyle, J. L. (1990). "Isolation of plant DNA from fresh tissue." *Focus*, 12:13-15.
Durand, R. and Durand, B. (1990). "Sexual determination and sexual differentiation." *Critical Review in Plant Sciences*, 9:295-316.
Faeti, V., Mandolino, G., and Ranalli, P. (1996). "Genetic diversity of *Cannabis sativa* germplasm based on RAPD markers." *Plant Breeding*, 115:367-370.
Gamborg, O. L. (1966). "Aromatic metabolism in plants. II. Enzymes of the shikimate pathway in suspension cultures of plant cells." *Canadian Journal of Biochemistry*, 44:791-799.
Hartsel, S. C., Loh, W. H. Y., and Robertson, L. W. (1983). "Biotransformation of cannabidiol to cannabielsoin by suspension cultures of *Cannabis sativa* L. and *Saccharum officinalis* L." *Planta Medica*, 48:17-19.

Helentjaris, T., King, G., Slocum, M., Siederstrang, C., and Wegman, S. (1985). "Restriction fragment polymorphisms as probes for plant diversity and their development as tools for applied plant breeding." *Plant Molecular Biology,* 5:109-118.

Heitrich, A. and Binder, M. (1982). "Identification of (3R,4R)-$\Delta^{1(6)}$-tetrahydrocannabinol as an isolation artifact of cannabinoid acids formed by callus cultures." *Experientia,* 38:898-899.

Hemphill, J. K., Turner, J. C., and Mahlberg, P. G. (1978). "Studies on growth and cannabinoid composition of callus derived from different strains of *Cannabis sativa.*" *Lloydia,* 41:453-462.

Hirata, K. (1924). "Cytological basis of the sex determination in *Cannabis sativa* L.." *Japanese Journal of Genetics,* 4:198-201.

Hormaza, J. I., Dollo, L., and Polito, V. S. (1994). "Identification of a RAPD marker linked to sex determination in *Pistacia vera* using bulked segregant analysis." *Theoretical and Applied Genetics,* 89:9-13.

Hu, C. Y. and Wang, P. J. (1983). "Meristem, shoot tip, and bud cultures." In Evans, D. A., Sharp, W. R., Ammirato, P. V., and Yamada, Y., (Eds.), *Handbook of plant cell culture,* Volume 1, pp. 117-227. New York: MacMillan.

Jones, A. and Veliky, I. A. (1980). "Growth of plant cell suspension cultures on glycerol as a sole source of carbon and energy." *Canadian Journal of Botany,* 58:648-657.

Klimyuk, V. I., Carroll, B. J., Thomas, C. M., and Jones, J. D. G. (1993). "Alkali treatment for rapid preparation of plant material for reliable PCR analysis." *The Plant Journal,* 3:493-494.

Knapp, S. J. (1994). "Mapping quantitative trait loci." In R. L. Phillips and I. K. Vasil (Eds.), *DNA-based markers in plants,* pp. 58-96. Dordrecht: Kluwer Academic Publishers.

Lebel-Hardenack, S. and Grant, S. R. (1997). "Genetics of sex determination in flowering plants." *Trends in Plant Sciences,* 2:130-136.

Loh, W. H. T., Hartsel, S. C. and Robertson, L. W. (1983). "Tissue culture of *Cannabis sativa* L. and *in vitro* biotransformation of phenolics." *Zeitschrift für Pflanzen physiologie,* 111:395-400.

Mandolino, G., Faeti, V., Zottini, M., Moschella, A., and Ranalli, P. (1995). "Il miglioramento genetico della canapa: aspetti biotecnologici." *Sementi Elette,* 2:57-60.

Mandolino, G., Carboni, A., Forapani, S., Faeti, V., and Ranalli, P. (in press). "Identification of DNA markers linked to the male sex in dioecious hemp *(Cannabis sativa L.)*." *Theoretical and Applied Genetics.*

Meijer de, E. P. M. and van Soest, L. J. M. (1992). "The CPRO *Cannabis* germplasm collection." *Euphytica,* 62:201-211.

Mellor, F. C. and Stace-Smith, R. (1977). "Virus-free potatoes by tissue cultures." In J. Reinert and Y. P. S. Bajaj (Eds.), *Plant cell, tissue and organ culture,* pp. 616-636. Berlin: Springer-Verlag.

Miller, C. O. (1968). "Naturally-occurring citokinins." In F. Whigtman and G. Setterfield (Eds.), *Biochemistry and physiology of plant growth substances*, pp. 33-45. Ottawa: Runge Press.
Moham Ram, H. Y. and Nath, R. (1964). "The morphology and embriology of *Cannabis sativa* L." *Phytomorphology*, 14:414-429.
Moham Ram, H. Y. and Sett, R. (1985). "*Cannabis sativa*." In *Halevy Handbook of flowering plants*, Volume 2, p. 131. Boca Raton, FL: CRC Press.
Moschella, A., Mandolino, G., and Ranalli, P. (1996). "Condizioni che favoriscono la rigenerazione *in vitro* di *Cannabis sativa* L." Abstracts of the *40th Annual Meeting of the Italian Society of Plant Genetics*, Perugia, September 18-21:193.
Murashige, T. and Skoog, F. (1962). "A revised medium for rapid growth and bioassays with tobacco tissue cultures." *Physiologia Plantarum*, 15:473-497.
Pate, D. W. (1994). "Chemical ecology of *Cannabis*." *Journal of the International Hemp Association*, 1:29-37.
Pennazio, S. and Ridolfi, P. (1974). "Potato virus X eradication in cultured potato meristem tips." *Potato Research*, 17:333-335.
Peters, J. E., Wu, P. H. L., Sharp, W. R., and Paddock, E. F. (1974). "Rooting and the metabolism of nicotine in tobacco callus cultures." *Physiologia Plantarum*, 31:97-100.
Pillay, M. and Kenny, S. T. (1996). "Random amplified polymorphic DNA (RAPD) markers in hop, *Humulus lupulus*: Level of genetic variability and segregation in F_1 progeny." *Theoretical and Applied Genetics*, 92:334-339.
Restivo, F. M., Tassi, F., Biffi, R., Falavigna, A., Caporali, E., Carboni, A., Doldi, M. L., Spada, A., and Marziani, G.P. (1995). "Linkage arrangement of RFLP loci in progenies from crosses between double haploid *Asparagus officinalis* L. clones." *Theoretical and Applied Genetics*, 90:124-128.
Richez-Dumanois, C., Braut-Boucher, F., Cosson, L., and Paris, M. (1986). "Multiplication végétative *in vitro* du chanvre (*Cannabis sativa* L.). Application à la conservation des clones sélectionnés." *Agronomie*, 6(5):487-495.
Sakamoto, K., Shimomura, K., Komeda, Y., Kamada, H., and Satoh, S. (1995). A male-associated DNA sequence in a dioecious plant, *Cannabis sativa* L." *Plant Cell Physiology*, 36:1549-1554.
Shattuck-Eidens, D. M., Bell, R. N., Neuhausen, S. L., and Helentjaris, T. (1990). "DNA sequence variation within maize and melon: Observations from polymerase chain reaction amplification and direct sequencing." *Genetics*, 126:207-217.
Southern, E. M. (1975). "Detection of specific sequences among DNA fragments separated by gel electrophoresis."*Journal of Molecular Biology*, 98:503-517.
Staba, E. J. (1977). "Tissue culture and pharmacy." In J. Reinert and Y. P. S. Bajaj (Eds.), *Applied and fundamental aspects of plant cell, tissue and organ culture*, pp. 694-702. Berlin: Springer-Verlag.
Stace-Smith, R. and Mellor, F. C. (1968). "Eradication of PVX and PVS by thermotherapy and axillary bud culture." *Phytopathology*, 58:199-203.
Steward, F. C. (1983). "Reflections on aseptic culture." In D. A. Evans, W. R. Sharp, P. V. Ammirato, and Y. Yamada (Eds.), *Handbook of plant cell culture*, Volume 1, pp. 1-10. New York: MacMillan.

Tanskley, S. D., Young, N. D., Paterson, A. H., and Bonierbale, M. W. (1989). "RFLP mapping in plant breeding: New tools for an old science." *Bio/Technology,* 7:257-264.

Veliky, I. A. and Genest, K. (1972). "Growth and metabolites of *Cannabis sativa* cell suspension cultures." *Lloydia,* 35:450-456.

Warmke, H. E. and Davidson, H. (1944). "Polyploid investigation." *Yearbook of the Carnegie Institution of Washington,* 43:135-139.

Williams, J. G. K., Kubelik, A. R., Livak, K. J., Rafalski, J. A., and Tingey, S. V. (1990). "DNA polymorphisms amplified by arbitrary primers are useful as genetic markers." *Nucleic Acids Research,* 18:6531-6535.

Yamada, I. (1943). "The sex chromosomes of *Cannabis sativa* L." *Reports of the Kihara Institute of Biological Research,* 2:64-68.

Yamada, Y. (1984). "Selection of cell lines for high yields of secondary metabolites." In I. K. Vasil (Ed.), *Cell culture and somatic cell genetics of plants,* Volume 1, pp. 629-636. Orlando, FL: Academic Press.

Zottini, M., Moschella, A., Mandolino, G., and Ranalli, P. (1996). "Efforts to improve tissue culture in hemp (*Cannabis sativa* L.)." *Abstracts of the Third European Symposium on Industrial Crops and Products,* Reims, France, April 22-24:15.

Chapter 10

Alkaline Pulping of Fiber Hemp

Birgitte de Groot
Gerrit J. van Roekel Jr.
Jan E. G. van Dam

INTRODUCTION

Fiber Hemp in the Netherlands

Nonfood fiber crop growing next to potatoes, wheat, and sugarbeets

In the early 1980s, a group of farmers in northeast Netherlands (the area of the Veenkoloniën) and students of the Agricultural University Wageningen carried out a prefeasibility study on the possibilities to introduce a "fourth crop." In this area about 50 percent of the farmland is used for potatoes (for starch production), about 25 percent for wheat, and about 25 percent for sugar beets. A new crop should be profitable, broaden the rotation scheme, and reduce the need for pesticides and herbicides. This study led to the rediscovery of fiber hemp in the Netherlands. It was selected as the most promising "fourth crop," as a nonfood, easy-growing crop for farmers, providing the pulp and paper industry with domestic fibers. A range of preliminary trials on growing, harvesting, ensilage, fiber cleaning and separation, and pulp and paper processing and economical viability were carried out. The promising results persuaded regional and national governments to support a larger research project. The Dutch hemp project was started in 1990 at a number of agricultural research institutes (DLO) and at the mentioned university in Wageningen. The Dutch paper and board industry showed some interest in the research results and offered advice and practical assistance. The results of the research project were promising, and the next step in the feasibility study for pulping hemp fibers for paper purposes will be a demonstration plant.

Fiber Makeup of Paper

Paper products are made from a mix of virgin pulps, recycled pulps, and additives, the specific mix depending on the paper's end use (varying from massive board and sanitary papers to currency paper grades). The total fiber mix for paper in Europe consists of about 56 percent wood pulp and 44 percent recycled fibers (Matussek, Pappens, and Cenny, 1996).

About 11 percent of the world's virgin pulp is made from nonwood fiber: straw (46 percent), bagasse (14 percent) and bamboo (6 percent) being the most important nonwood fiber sources (Atchison, 1996). The largest nonwood pulping nation is China, producing 74 percent of the total nonwood pulp. Although the United States, Canada, and most European countries use little nonwood pulp, the Chinese pulp mix consists of about 60 percent nonwood pulp, about 25 percent recycled fibers, and less than 15 percent wood pulp (Atchison, 1996). The world hemp paper pulp production is estimated at 120,000 tons/year (FAO, 1991), amounting to only 0.05 percent of the total paper production. Hemp pulps are generally blended with other (wood) pulps for specialty paper production. There is (at this moment) no significant production of 100 percent true hemp paper.

Potential Pulp Markets for Hemp Fibers

Table 10.1 summarizes the potential markets for hemp bast and hemp woody core fibers. The bast fibers are very valuable, but the specialty paper (e.g., banknotes, tea bags, cigarette paper) market is relatively small. Any new bulk production volume of specialty pulp in excess of 20,000 tons yearly would destroy the world market. Large scale fiber hemp cultivation can therefore only be viable when the produced pulp can compete on the wood pulp market for bulk applications (linerboard, tissue, hygienic, printing and writing paper grades).

Paper applications of fiber hemp are discussed in this chapter. Nonetheless, many other market outlets are also being explored, particularly in (geo)textiles (as nonwovens) and fiber reinforced composite and building materials. Researchers at the Daimler-Benz car company

Table 10.1. Paper market potential of fiber hemp pulp.

fiber hemp	%	comparable pulp/cellulose	purpose	Dutch market potential T/year
bast fibers	35	cotton linters cellulose, abaca cellulose	thin, strong, and durable specialty papers	3,000
		softwood CTMP (chemi-thermo-mechanical pulp), softwood cellulose	strength in testliner, LWC (light weight coated), sanitary	60,000
woody core	65	hardwood CTMP (aspen)	smoothness and printability in board and coated grades	130,000
		hardwood cellulose (birch, eucalyptus, mixed hardwoods)	smoothness and printability in printing and writing grades	560,000

Source: van Kemenade et al. (1993). "Starting points and options for a Dutch demonstration pulp mill." In J. M. van Berlo (Ed.), *Paper out of hemp from Dutch soil* (Dutch), pp. 184-207. Wageningen, Netherlands, ATO-DLO.

have found that fibre hemp can be used instead of glass fiber (which causes environmental damage after disposal) to reinforce plastic components in vehicles (Anonymous, 1996). They claim that fiber hemp is more economical than flax, and matches or surpasses flax in terms of performance potential (hemp fibers are more rigid than flax fibers).

Since the market outlet for paper purposes for hardwood type fibers is large and still growing, hemp woody core pulp might be developed to fit this market.

BAST FIBER PULPS FOR PAPER APPLICATIONS

Hemp bast fibers as reinforcing component in paper and board grades

Uses of Bast Fibers

Like other plants offering long and strong fibrous material (jute, kenaf, flax) hemp has traditionally been regarded as material for textiles, sailing and fishing gear, and for strong, thin, durable specialty papers. The outer parts of the hemp stem consist of long bast fibers (5 to 50 mm, with an average fiber length of 16 mm).

At present twenty-three paper mills use hemp fiber, at an estimated world production volume of 120,000 tons/year (FAO, 1991). Most of the mills are located in China and India, and produce moderate quality printing and writing paper. Typically, these mills do not have a fixed source of fiber, using whatever can be found in the region.

About ten mills, located in the United States, Europe, and Turkey, produce so-called specialty papers, used in applications with high tear strength and high wet strength that can generally only be produced from hemp, flax, cotton, and other nonwood fibers:

- cigarette paper: popular American cigarette brands contain 50 percent hemp fibers in cigarette paper and filters. Some countries prescribe the use of hemp in cigarette paper by legislation, because wood fibers such as spruce generate hazardous fumes when incinerated;
- security, Bible, greaseproof, and various specialty art papers;
- insulating papers (for electrical condensers); and
- specialty nonwovens, filter paper (for technical and scientific uses), coffee filters, tea bags

Classical Hemp (Bast Fiber) Pulping

The average hemp pulp and paper mill produces approximately 5,000 tons/year. This is very little, compared to the present-day world scale wood pulp mill at 1 million tons a year.

The existing hemp mills can still produce at this extremely small volume, because of their specialty products. This partly explains the high price for a hemp bast fiber pulp of about 3,100/t U.S. dollars versus 650 U.S. dollars for bleached softwood pulp market prices for May, 1997.

Hemp mills in the United States, Europe, and Turkey are, because of their small size and archaic technology, unable to cope with environmental legislation. Some mills survive by shipping their waste water to a large wood pulp mill nearby; others will have to close down. There is a shift in capacity toward countries that do not yet give environmental issues priority.

Most mills mainly process the long hemp bast fibers, which arrive as baled and cleaned ribbon at preprocessing plants located near the cultivation areas. The bales are opened and fed into a spherical tank, the digester. Water is added (4:1 liquor:fiber ratio), together with chemicals (usually sodium hydroxide and sodium sulphide) to remove lignin and pectin (coloring and sticky components). The fibers are cooked (up to eight hours) at elevated temperature and pressure, until the fibers are separated from each other. After cooking, the used chemicals and degraded and extracted components are separated from the fibers by thorough washing. Here, a polluting waste emerges from the process, which is often discharged into the local surface water; but chemicals can be recovered (at high equipment costs). The cleaned fibers are fed into a Hollander beater in which the fibers are circulated for up to twelve hours per batch. During the beating process, the fibers are cut to the right length and attain the required surface roughness, thus providing the required bonding capacity. Some mills add bleaching chemicals in the beater; other mills pass the pulp from the beating machines to separate tanks for bleaching. These separate bleaching treatments often use chlorine compounds, which are also discharged into the environment. The bleached pulp can either be pumped to the paper machine, or pressed to a dryness suitable for transportation to a paper mill elsewhere.

As the process takes more than twenty hours and the throughput is low, the costs of equipment and handling per ton product are high.

Papermaking Properties of Hemp Bast Compared to Softwood Fibers

The bast fiber is distinguished from softwood fibers by a much greater length, a thicker cell wall, and a very small lumen. The fibers are rigid and not collapsible, even after considerable beating action. The hemp bast fiber is very high in cellulose content (see Table 10.2), and very low in lignin. Most of the 3 to 5 percent lignin is situated in the middle lamella, combined with pectin substances (van den Ent and Harsveld van der Veen, 1994). The low lignin content allows mechanical bast fiber pulps to be characterized as woodfree pulps. The thick hemp bast fiber walls imply low conformability, so there is not much

Table 10.2. Chemical composition of some fiber crops and wood species (on extracted dry wood basis).

	glucan	xylan	mannan	lignin
Picea abies (spruce)[1]	44.3	7.6	10.3	28.6
Pinus sylvestris (pine)[1]	44.8	7.2	7.6	27.8
Betula verrucosa (birch)[1]	37.5	24.6	0.5	19.5
Cannabis sativa (hemp) woody core bast fibers	37.7 66.7	16.7 1.5	1.2 1.9	22.1 4.0
Triticum vulgare (wheat)[2]	30.4	12.2	0.4	9.9

Note: [1]=Rydholm, 1985; [2]=Nordkvist, 1989.
Source: Sven A. Rydholm. (1985) (reprint ed.). *Pulping Processes*. Malabar, FL: R. E. Krieger Publishing Company; Erik Nordkvist, Hadden Graham, and Per Aman. (1989). "Soluble lignin complexes isolated from wheat straw *(Triticum arvense)* and red clover *(Trifolium pratense)*." *Jounal of the Science of Food and Agriculture*, 48(3): 311-321.

bonding to be expected. The tear strength is particularly high due to the high fiber length. The Dutch research for new applications of hemp fibers in bulk pulp and paper markets focuses on printing and writing grade papers, boards, and tissue or fluff applications. The extreme native fiber length of hemp bast pulp causes considerable problems in handling the pulp in present stock preparation systems and on modern paper machines for mentioned applications. Spinning problems in pumps, refiners and screens, and formation problems and knots forming on the paper machine do not allow the use of untreated hemp fiber. A new pulping process for hemp should therefore aim to produce fibers at an optimum length with maximum strength but without spinning and formation problems.

HEMP WOODY CORE AND HARDWOOD PULP

Progressive demand for hardwood pulp indicating opportunities for hemp woody core

Hemp Woody Core Related to Hardwood and to Nonwood Fibers

The inner parts of the hemp stem, known as woody core or shives, consist of fibers with a length of 0.5 to 0.6 mm (with an average of 0.55) and represent about 65 percent of the weight of the stem. This material is commonly discarded, or burned.

Botanically as well as chemically (see Tables 10.2 and 10.3), hemp woody core is comparable to hardwood. In this respect hemp is unlike most annual fiber crops that must be treated as straw, requiring special effluent treatment, due to high silica content. Hemp is a dicotyledon and has a very low silica content, in contrast with monocotyledons such as straw or other grasses.

When eucalyptus pulp was introduced, papermakers were not impressed. The fiber length of 0.8 mm was 20 percent shorter than thought necessary. Although eucalyptus pulp was expected to fail, eucalyptus has now become a commercially important pulp. Similarly, hemp woody core (two-thirds of the stem weight) may be developed as valuable paper feed stock in the future, facilitated by continuously improved pulping and refining technology.

INTRODUCTION OF HARDWOOD AND RECYCLED FIBERS IN PAPER

Traditionally spruce and pine are used for chemical softwood pulp to produce strong, thick paper. Chemical hardwood pulp has been used increasingly for the last twenty-five years (Baker, 1995). The pulp and paper industry has employed various strategies to reduce the cost of fiber stock.

Table 10.3. Botanical classification of fiber hemp and other renewable fiber sources.

Subdivision	Class	Family	Species
Gymnospermae	Coniferae	Piceae	*Picea abies* (spruce)
			Pinus sylvestris (pine)
Angiospermae	Dicotyledonae	Betulaceae	*Betula verrucosa* (birch)
		Fagaceae	*Fagus sylvatica* (beech)
		Saliceae	*Populus tremuloides* (poplar)
		Cannabinaceae	*Cannabis sativa* (hemp)
		Urticaceae	*Boehmeria nivea* (ramie)
		Linaceae	*Linum usitatissimum* (flax)
	Monocotyledoneae	Gramineae	*Triticum vulgare* (wheat)
			Phylostachys puberula (bamboo)
			Saccharum officinarum (bagasse)

Source: de Groot, 1995, p. 256.

Larger use is made of faster growing trees, such as poplar and eucalyptus, to produce thinner paper. Also, recycled paper is increasingly used as fiber stock. These strategies require upgrading: using improved technologies to produce better and better paper from cheaper and cheaper fiber stocks. Most market pulps in Europe are produced in Sweden, Finland, and Portugal, while the net imported pulp to Europe (about 23 percent of the used market pulp) is mainly produced in Canada and the United States (Stefan, 1995). The amount of Western European market pulp is expected to decline due to higher levels of integration (producing paper instead of selling pulp [Stefan, 1995]).

For some paper mills this may be a reason to invest in pulp mills for domestic fiber sources such as fast-growing hardwoods and annual fiber crops such as hemp.

The experience, skills, and knowledge in the papermaking industry in handling hardwood and recycled fibers may prove very valuable for the design of processes for agricultural fiber crops, as both accurate cleaning and screening, and adapted pulping processes have to be developed for pulp production from agricultural fiber crops.

Upgrading Hardwood Pulp

Traditionally used as filler pulps, hardwood pulps have become important paper furnishers (fine papers may contain 70 to 90 percent hardwood [Baker, 1995]). To use hardwoods to their maximum potential, refining processes must be gentle, avoiding cutting (adapted refiner plate profiles), and using medium consistency refining. Mechanical pulping processes (used for fiber separation and fibrillation) also shorten the fibers, which is no problem for softwood fibers (with a length of about 3 mm), but is more damaging for hardwood fibers of 1 mm and less. This explains why no thermomechanical processes are presently used for hardwood fibers. Currently, processes developed for high-yield hardwood (HYH) pulping (with limited use of chemicals) are successfully used in Canada (Ford and Sharman, 1996; Meadows, 1996). For example, the produced hardwood pulps are used as ingredients in lightweight coated papers in European mills.

Hardwood pulps can yield a well-formed strong sheet, to a point where softwood and hardwood pulps give nearly identical properties (Baker, 1995). Although worldwide, bleached softwood kraft is still the dominant market pulp grade (17 million tons were produced in 1994), the importance of bleached hardwood kraft is still increasing (12.9 million tons were produced in 1994, a rise with 13 percent compared to 1993 [Stefan, 1995]). Printing and writing grade papers in the Netherlands consist of nearly equal amounts of hardwood and softwood fibers. Prices of bleached (long-fibered) softwood grades are about 10 percent higher than prices of bleached (short-fibered) hardwood grades.

Prospects for Hardwood Pulp Production

Indonesia is building two new bleached hardwood kraft mills and its total pulping capacity will rise to 1.5 million tons (which is consistent with the expected increase in pulp demand in Southeast Asia). Furthermore, increasing pulp demands in South America will be met by installing new bleached kraft mills for hardwood (0.5 million tons) in Brazil, while a project for another 0.5 million tons for eucalyptus pulp production is being studied in Chile (Stefan, 1995).

Meanwhile, as has been pointed out, high-yield hardwood pulp has been introduced successfully as market pulp, indicating that high-yield pulping might be an option for hemp fibers.

IMPORTANT PULPING PROCESSES AND THEIR SIGNIFICANCE FOR HEMP

Kraft Pulping: From Odorous Process to Low Pollution Pulping Systems

The kraft process is the major pulping process, responsible for about 90 percent of the world production of lignin-free pulp (Matussek, Pappens, and Cenny, 1996). The kraft process separates and purifies fibers with a solution of sodium sulphide and sodium hydroxide, at a reaction temperature of 170°C. The conventional kraft cooking process, combined with a chlorine-based bleaching system, is malodorous and polluting. However, environmental concerns of paper consumers, as well as governmental restrictions, have urged pulp mills to develop less polluting pulping and bleaching technologies. The modernized kraft process is used in developed, environmentally conscious countries (Ahlgren, 1991; Anonymous, 1997).

Nonwood Processing: New or Adapted Processes and Equipment

Modern kraft mills produce up to 750,000 tons of dry pulp yearly (minimizing investment and chemical recovery costs on ton product

[Stefan, 1995]). With ten tons dry stems/ha, and yields (equaling cellulose contents) of about 70 percent for hemp bast fibers (35 percent of the stem) and 40 percent for hemp woody core fibers, 1 hectare yields five tons of cellulose. About 100,000 hectares should be needed yearly for a mill with a production of 500,000 tons/year. Consequently, large logistic problems must be solved (storage, transportation, guaranteed annual supply) and large investments must be made (apart from the start-up costs), before such a mill can be built for kraft pulp production using fiber hemp or any other annual fiber crop.

Adapted pulping processes have been described, mainly for straw or other monocotyledons, to reduce the economically necessary production scale. For monocotyledons, the silica in the effluent complicates recausticizing, by which reaction chemicals are recovered.

Sodium may be replaced by potassium, using process effluent as fertilizer (avoiding high recovery costs) for remunerative small-scale mills (Wong, 1994; Sameshima and Ohtani, 1995). Also, pulping without silica extraction (retaining silica in the pulp), as is used for the Milox process (using formic and peroxyformic acid), is an option (Seisto and Poppius-Levlin, 1995).

Furthermore, processes using alkaline chemicals such as borate and phosphate, that enable the use of autocausticizing recovery systems have been developed to reduce costs of recovery equipment (Janson 1980, 1992; Janson et al., 1994).

Silica is not a problem for hemp pulping, but could be a problem if other annual fiber crops should also be processed in the same pulp mill. Fertilizer-producing processes are rejected for most Western European countries, since there is no shortage of fertilizer. However, the other systems (autocausticizing process, Milox process) may prove worthwhile, if hemp is to be used for lignin-free (low-yield) pulping.

Pulping systems using ethanol enable separating lignin (degradation products) from carbohydrate degradation products in the effluent. These systems produce not only cellulose, but also lignin and hemicellulose products. However, the present market for lignin products is very small. Another strategy to reduce the minimum mill scale is to process hemp similarly as with the kraft process, but without sulphide. In general, nonsulphur, continuous NaOH pulping can be executed with cheaper recovery, as furnace temperatures can be lower than needed to reduce sodium sulphate to sodium sulphide.

If no monocotyledons are used (e.g., only hemp, flax, and hardwoods, with low silica content), similar recovery systems as developed for high-yield alkaline sodium peroxide pulping can be used. These installations have also been developed to provide modular recovery expansion for kraft mills (saving the costs of downtime required for rebuilding existing recovery boilers) and can be built with a capacity of 225 tons/day of dry solids (Lecsek, 1994; Meadows, 1996).

New Developments: High-Yield Hardwood (HYH) Pulping Processes

During the last ten years, high-yield (85 percent) processes have been developed for hardwood that can produce pulps with similar properties such as hardwood kraft pulp, at significantly lower operating and capital cost (Ford and Sharman, 1996). High-yield processes maximize the use of fiber resources, use only oxygen-based chemistry for bleaching, and can be designed for low water use or with a zero liquid effluent installation. The simple process technology allows the economical construction of relatively small plants (50 to 200,000 tons/year). It has been suggested that HYH mills in Europe could be installed in areas where appropriate fiber sources are available, such as in Spain and Portugal (using eucalyptus) and in Scandinavia, Russia, and the Baltic States (using birch). Hemp woody core, when available, could be pulped similarly.

When a HYH pulp mill is integrated with a paper mill, production costs can be $30/ton lower than for hardwood kraft pulp. Comparing nonintegrated mills, HYH pulp can be produced $10/ton cheaper than hardwood kraft pulp.

ALKALINE PULPING OF HEMP

Hemp Bast Fibers

Introduction

The traditional chemical pulping process for fiber hemp involves excessive beating. This reduces the average fiber length to less than 2 mm, yet some remaining tough and long fibers can cause spinning or knotting problems. Apart from this technical problem, the intensive handling and effluent treatment costs for a traditional small mill result in a price level that is three to four times the level of wood pulps. In order to supply hemp bast pulp to bulk markets as was proposed in the Dutch hemp program, alternative technology was tested at ATO-DLO. (More details can be found in van Roekel 1995, 1996; van Roekel et al., 1995.)

Mechanical Extrusion Pulping Procedures for Hemp Bast Fibers

The basic principles of extrusion pulping have been developed during the last two decades in France, with successful results in pilot and industrial applications processing annual plants such as cotton, hemp, and flax. Extrusion pulping can be a mechanical or chemi-mechanical pulping method. It has been proven particularly

Figure 10.1. Diagram of the experimental setup at ATO-DLO.

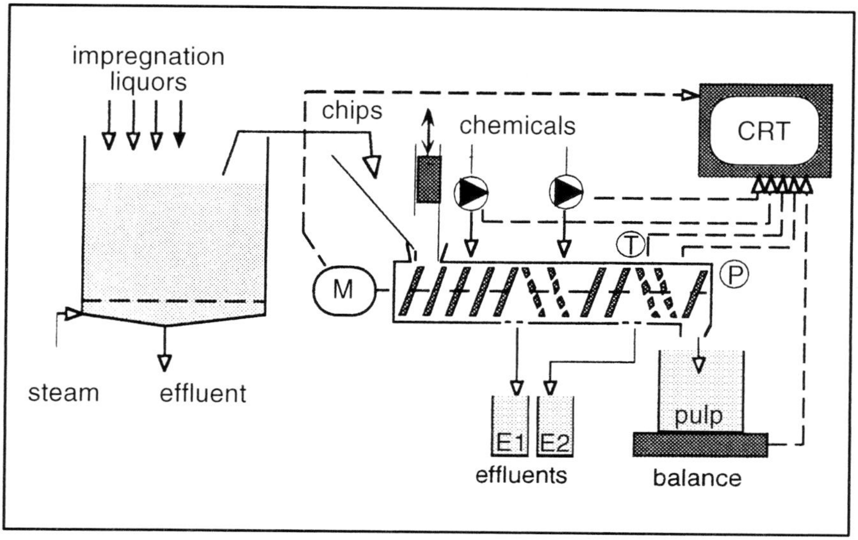

Source: ATO-DLO, G. T. van Roekel; van Roekel et al. (1995); van Roekel (1996).

useful for very long vegetable fiber materials. Presently, some twenty industrial installations are operational in pulping cotton, flax, and hemp. Mechanical extrusion pulping of hemp bast fibers produces pulps with the desired fiber length distribution. These pulps, from fibers with only 3 to 5 percent lignin, can be characterized as woodfree, and can therefore be applied in woodfree printing and writing paper grades.

The pulping extruder consists of a "tilted eight-shaped" barrel (∞) with two corotating intermeshing screws inside. Fibrous material is fed into the barrel by transport screws.

Farther down the barrel, a screw with its flight pitch reversed (reverse screw element or RSE) causes severe compression of the fiber mass. The generated pressure forces the fiber mass to flow through slots machined in the RSE flights, where a high shear field causes defibration and cutting of the fibers. Excess water pressed out of the fiber mass is extracted through barrel filters placed upstream from the RSE. The pressure drop created in passing a RSE heats the pulp mass and provides rapid impregnation of liquids supplied through an injection port downstream from the RSE. The combination of transport screw, reverse screw filter, and injection port constitutes one defibration zone. A pulping extruder can be set up to hold four separate defibration zones. The pressurized fiber mass in the RSE separates the liquids in the subsequent zones from each other. It is thus possible to create a sequence of different operations within one machine treatment. Additional heating of the pulp is provided by steam injection.

The original BiVis (French for twin-screw) process was designed with two pulping extruders in series, the first for impregnation and partial cutting of fibers, the second for bleaching and additional cutting. Often intermediate retention chests are added to extend the reaction time for pulping or bleaching chemicals.

The extrusion process for hemp bast pulps evaluated at ATO-DLO was derived from the BiVis process, but uses only one extruder, followed by a disk refiner. Further, the fibrous material is subjected to cold alkaline impregnation and subsequent washing before entering the extrusion process.

A number of pulping experiments were carried out to evaluate the technical feasibility of using hemp pulps for applications in bulk paper markets. The following pulps were produced: unbleached mechanical pulp (XP) and unbleached alkaline-mechanical pulp (AXP) both for linerboard and food board, and bleached alkaline-peroxide mechanical pulp (APXP) for printing and writing grade paper. The following section describes the general experimental procedures for production of these pulps on a laboratory scale and the main test results.

Hemp bast fiber preparation. The *Cannabis sativa* variety Kompolti Sárgaszárú was grown locally and harvested by the end of August 1990. The dry and clean hemp was separated by breaking and scutching to 2 percent remaining core fragments in the bast fiber fraction. The cleaned bast fiber lints were cut to 20 mm lengths with a guillotine cutter to improve handling. The cut bast lints are referred to as chips.

Impregnation and preheating. Prior to extrusion pulping, bast fiber chips were impregnated and preheated with saturated steam at atmospheric pressure. For alkaline impregnation a sodium hydroxide solution was used for immersion of the chips. After draining, the fiber mass was washed by percolation with cold water for thirty minutes. Then the chips were preheated to $100°C$ by atmospheric steaming.

The impregnation yield drops to 92 percent once alkali is added to the cold impregnation water. It is assumed that mainly extractable components are removed in the impregnation stage.

Extrusion pulping. The laboratory extruder could only hold two defibration zones, so the fiber mass was passed through the extruder twice to simulate an industrial scale extruder. The intermediate pulp was reheated before the second run.

At a comparable pulp throughput, the pulp consistency of XP is relatively high. This effect is explained by the higher shear compression caused by a less flexible fiber as compared to the alkaline pretreated pulps.

The specific power consumption decreases significantly with alkaline pretreatment of the chips. Under alkaline conditions, the impregnation effluent tends to gelatinize, suggesting the presence of carbohydrate polymers. The pulp yields of alkaline pretreated pulps are significantly lower than for the untreated pulp, probably due to enhanced extraction.

Bleaching. Optional alkaline peroxide bleaching of the pulp was performed in conjunction with the two extrusion pulping treatments by injecting bleaching liquors into the extruder. The bleaching system is similar to the alkaline peroxide mechanical pulping (APMP) process for wood chips (Bohn and Sferrazza, 1989; Cort and Bohn, 1991). After each extruder passage the pulp was allowed to react for 1.5 hours at 70°C. Then, the pulp was washed by percolation with cold water.

The use of a tighter screw configuration, designed to improve extraction of bleaching liquor in the second extrusion stage, also resulted in increased power consumption (increased pressing action).

Refining. The extruded pulps were further refined in a Bauer CP-12 pressurized laboratory refiner setup.

Pulp and Paper Characteristics

Fiber length. The weight weighted average fiber length of 2.5 mm of APXP at 19°SR (using the Kajaani fiber length analysis) indicates that spinning or formation problems are not to be expected.

Beating and strength properties. The beating degree of a typical XP with two subsequent TMP (thermo-mechanical pulp) treatments at 1.5 bar (128°C) increases relatively fast at a relatively low energy consumption as compared to a typical spruce TMP. The tensile and burst strengths develop with beating to a moderate level compared to spruce TMP. The pulp density is higher, due to the higher coarseness of the fibers. The tear strength slowly decreases with beating, showing continued fiber length reduction. The optical properties can be compared to softwood TMP.

The energy consumption of a typical AXP with two subsequent TMP treatments at 1.5 bar is significantly lower than of the XP, and about half of a typical spruce CTMP (chemi-thermomechanical pulp). The alkaline pretreatment does not cause a lower beating degree. Taking the degree of beating into account, the tear, tensile, and burst strengths have increased significantly. Compared to spruce CTMP the strength properties are comparable, but at a much lower energy consumption. The density of hemp AXP is relatively high due to the fiber coarseness. The optical properties are comparable to spruce CTMP, taking into account that no sulphur was used.

APXP (bleached pulp) was subjected to an atmospheric treatment at 100°C (RMP: refiner mechanical pulping) and a pressurized treatment at 128°C (TMP). The differences between the TMP and RMP treatments of APXP are not significant, except for the energy consumption. It appears that a RMP treatment can induce strength properties comparable to the TMP treatment. The energy consumption of APXP is higher than for AXP. The pulp density of APXP is higher than for AXP, and is considerably higher than for a spruce APMP pulp, but still lower than a NBSK (northern bleached softwood kraft) pulp. Tensile and burst strengths of APXP and AXP are at the same level. The optical properties achieved with 3.5 percent peroxide are satisfactory. The produced hemp bast fiber pulps have been evaluated for use in various bulk market applications, of which two examples are presented.

Hemp bast AXP for testliner reinforcement. In the production of testliner, a considerable amount of preconsumer kraft waste is utilized for reinforcement of the bulk waste paper raw material. Presently, the most important properties for reinforcement of linerboard are tear strength and SCT (short-span compressive test) stiffness. Improvement of the tear factor of testliner for product quality improvement would require about one-third of the amount of kraft waste. However, the input of hemp AXP is limited by the allowed decrease in SCT stiffness when replacing kraft waste as a reinforcer.

Hemp bast APP for printing/writing grade paper. Hemp bast APXP can technically be characterized as a woodfree (chemical) pulp because of the original low lignin content. Hemp bast APXP (RMP) was compared to both a typical NBSK and a typical BCTMP (bleached chemi-thermomechanical pulp). The APXP density is better than that of NBSK, but not as good as BCTMP. The tensile strength is inferior compared to NBSK, but comparable to BCTMP. The burst strength is comparable, probably due to the high tear strength of APXP. The brightness is on the level of softwood BCTMP.

Conclusions

It was found that the best pulping method was alkaline (peroxide) extrusion pulping, at 80 percent yield and with 50 percent of the energy requirements of mechanical softwood pulps. Peroxide can be added for higher brightness. The long fibers are cut, preventing spinning and knotting problems in modern stock and paper machines, while sufficient strength is retained to be used as reinforcing pulp for testliner or printing and writing grades.

The (unbleached) AXP properties are sufficient to consider application as a reinforcement pulp for testliner production and for replacing preconsumer kraft waste. The (unbleached) XP compares to AXP, but shows higher energy consumption with a slower dewatering rate, and has better tear strength but higher density than softwood TMP. The properties of bleached hemp bast APXP are sufficient to consider application as replacement of softwood BCTMP in printing and writing grade papers. APXP may also replace NBSK, although a lower tensile should then be accepted, while density and tear are significantly better. Considering the low lignin content, APXP can be regarded as a woodfree total chlorine-free bleached pulp.

Hemp Woody Core

Introduction

Although hemp bast fibers have been used for paper since the invention of pulping processes for papermaking (105 A.D.) in China, hemp woody core fibers are still little used for this purpose. However, some work has been done in the United States (Dewey and Merrill, 1916), Germany (Wedekind, Grasshof, and Müller, 1937; Wedekind, 1938), and more recently in Italy (Bosia, 1975; Bosia and Nisi, 1978), indicating that hemp woody core could be processed and used similarly as hardwood fibers. These experiments and our own prefeasibility studies (de Groot and Harsveld van der Veen, 1988; de Groot, van Zuilichem, and van der Zwan, 1988) showed that hemp woody core was a promising raw material for papermaking and that alkaline pulping was a potential pulping process for hemp woody core.

Although hemp woody core might be pulped together with the bast fibers, it was preferable to first study pulping of the material separately (because of the different nature of the fiber fractions), to explore the maximum potential of both fiber fractions.

These were the starting points for more fundamental research carried out at Wageningen Agricultural University and ATO-DLO; a summary of the results is presented hereafter.

Alkaline Swelling

In general, sodium hydroxide promotes fiber swelling and fiber flexibility. This suppresses cross-fiber fragmentation during mechanical treatment and promotes fibrillation and formation of interfiber bonding during papermaking, resulting in good mechanical paper properties. Several motives induced the study of alkaline swelling of hemp woody core chips. First, as it became clear that high-yield pulping might be a potential process for hemp woody core, the effects of alkaline peroxide mechanical pulping (APMP) on pulp yield, chemical composition, and swelling of hemp woody core chips were determined (de Groot et al., 1997). Second, it was clear that a fair amount of xylan and lignin were extracted, or degraded and extracted during the short heating-up period, before reaction temperatures of 150°C and higher were reached for alkaline pulping (de Groot, van Dam, and van't Riet, 1995). Furthermore, to obtain a homogeneous distribution of the cooking liquor at the reaction sites at the start of the process, hemp shavings were impregnated with cooking liquor, applying vacuum to remove air entrapped in the shavings. A preliminary impregnation experiment proved that the amount of material extracted from the shavings increased with the NaOH concentration used.

The study of alkaline swelling quantified the effects of NaOH concentration on yield losses at room temperature and at elevated temperature (de Groot et al., 1997). For hemp woody core chips higher swelling was found with demineralized water at 70°C, than with any of the used NaOH concentrations (varying from 0.06 to 2.3 mol/l).

Some alkaline lignin softening and extraction may be necessary to separate the individual fibers at the middle lamella and enhance fibrillation, to preserve fiber strength and create bonding potential of individual fibers necessary for papermaking. As pulp yield diminishes with increased NaOH concentration, this concentration should be minimized to limit yield losses.

In general, peroxide is used to enhance brightness. However, more lignin is also removed from hemp woody core chips than without peroxide treatment, the amount increasing with NaOH concentration.

Consequently, further optimization of the APMP process is necessary, on a larger scale, using the appropriate equipment, minimizing xylan and yield losses as well as mechanical strength losses, and enhancing sufficient fiber bonding and brightness.

Influence of Chip Thickness on Alkaline Delignification

Research on the impact of chip size, controlling the effects of impregnation and diffusion on homogeneous pulping, has led to more accurate chip thickness screening and modified impregnation in modern kraft mills, using softwood and hardwood chips (Pulliam, 1995; Meadows, 1995). The importance of diffusion effects on pulping of well-impregnated woody core chips was checked, using chips of different thicknesses.

We found that delignification and pulp yield of hemp woody core are not related to chip thickness. This may be the result of the relatively low density of hemp woody core; rendering diffusion of reaction chemicals and degradation products easier (than in softwood and hardwood chips), provided that the chips are well impregnated with cooking liquor. Also, the maximal thickness of hemp chips (derived from the dimensions of a hemp stem) is relatively low.

Monitoring Lignin Removal

Suggestions that lignin degradation can be monitored with UV-detection of the effluent at 280 nm (Dolk et al., 1983; Tikka and Virkola, 1986; Paulonis and Krishnagopalan, 1988; Trinh, 1988) led to trials comparing UV-absorption with lignin content in the effluent. Successful lignin measurement in the effluent would reduce the number of delignification experiments substantially.

For hemp woody core it was found that the UV-absorption at 280 nm in the effluent varied too much (more than 100 percent) during a pulping run, to simply assume that UV-absorption of effluent samples is linearly correlated with lignin removal. It would take a separate study of hemp woody core lignin and its behavior during alkaline pulping and related to process conditions, to establish the UV-absorption during pulping. This variation in absorption may be related to varying syringylpropane:guaiacylpropane ratios, amounts of p-hydroxybenzoate groups and amounts of conjugated carbonyl groups in the removed lignin fractions (Sarkanen, Chang, and Allan, 1967b). Thus, many experiments were carried out and each resulting pulp sample was analyzed (to determine remaining lignin content), to get data to calculate delignification kinetics of hemp.

However, for kraft delignification of softwoods (with lignin that is more homogeneous and essentially composed of gualacylpropane units [Sarkanen, Chang, and Allan, 1967a]), UV-detection of effluent seems satisfying (Vanchinathan and Krishnagopalan, 1995, 1997). Alternatively, a fluorescence method for on-line delignification control (real time measurement in undiluted pulp) has been patented (Jeffers and Malito, 1996), promising accurate and fast lignin measurement.

Alkaline Delignification

Treatment with sodium hydroxide, at temperatures around 170° C, promotes delignification (removal and degradation of lignin). This facilitates disintegration of wood in fibrous components and eliminates the coloring substances. Delignification and carbohydrate degradation were studied to detect how hemp woody core is affected by process conditions (temperature, time, NaOH-concentration) and to enable optimization of the process (de Groot et al., 1994, 1995). Delignification trials were started, first (preliminary trials) in small batch digesters, later in a specially designed flow-through reactor.

A large amount of hemicellulose is removed during the initial delignification stage; a small amount of lignin is removed. During the bulk delignification stage most of the lignin is removed; the remaining lignin can be removed during the rest of the delignification stage, accompanied by cellulose degradation. These reaction stages can be modeled with an integral calculation method, regarding consecutive reaction stages as the result of simultaneous reactions, depending on reaction time, temperature, and sodium hydroxide concentration. The results have been used to design a new model describing alkaline delignification and hemicellulose degradation in hemp woody core. We also applied this approach successfully on literature results on delignification of softwood and hardwood species.

Comparing delignification kinetics with literature values for softwood and hardwood, it is apparent that delignification of hemp is faster than for softwood and hardwood species.

Paper Qualities

In general, fiber bonding and fiber length are influencing paper strength. Fiber bonding is determining tensile and burst strength, while fiber length is important for tear strength (Rydholm, 1985; Strand, Saltin, and Bardwell, 1994; Saltin and Strand, 1992; Seth and Page, 1988; Page, 1994). Such physical treatments as beating generate fibers with fibrillated surfaces and produce more flexible fibers, conforming more easily to the shape of neighboring fibers. This results in enhanced interfiber bonding, improving wet-web, tensile and burst strength, and diminishing paper bulk (Atack, 1978; Iyengar, 1982; Rydholm, 1985; Jiménez, López, and Martínez, 1993; Dasgupta, 1994; Broderick et al., 1996).

Paper surface properties such as scattering (ability of a paper surface to scatter incoming light) and opacity (degree to which the fiber surfaces in the paper sheet scatter incoming light, thus preventing light transmission through the sheet) are determined by unbound surface (Clark, 1985; Rydholm, 1985; Strand, Saltin, and Bardwell, 1994). Increased interfiber bonds result in less unbound surface and less scattering (Stratton, 1991), while opacity also diminishes and the paper sheet becomes more transparent as fibers merge into one another (Rydholm, 1985). Opacity and scattering have been related to paper density, tensile and burst strength (Rydholm, 1985; Dasgupta, 1994).

We examined whether the described knowledge on wood pulps in relation to paper qualities applies to alkaline hemp woody core pulp and paper (de Groot et al., 1996). Furthermore, the papermaking potentials were established in relation to commercial straw and hardwood pulps.

Fiber length. The fiber length of unscreened, bleached pulp was measured with a Kajaani FS-200 optical analyzer. The pulp has a rather short weight averaged fiber length of 0.61 mm, 86.2 percent (weight averaged) of the fibers are longer than 0.25 mm, which is a little less than found for kraft hardwood cellulose (87 to 97 percent) and more than found for straw cellulose (82 percent). The short fiber length may contribute to lower tear strength (Page, 1994) and low wet-web strength (tensile and stretch), reducing the operational ability of the paper machine (Seth, 1995). However, in pulp mixes short fibers also improve mass uniformity and paper formation, contributing to wet-web properties (Kerekes and Schell, 1995) and improving printability in fine paper and printing paper grades (Baker, 1995). This suggests that adding a certain amount of hemp woody core fibers to a pulp mix may be beneficial for printing and writing purposes.

In general, pulping and beating affects hemp woody core pulp and paper in the same way as described for wood pulps.

Beating and strength properties. The pulps develop very easily, as their bulk is reduced from about 1.5 to 1.2 cm^3/g within only 500 PFI rotations, while normally for hardwood pulps 2,000 PFI rotations are required. In this respect the pulp is comparable with straw pulp, known to develop with a similar ease.

The measured tensile and burst values (respectively 6 to 9 km and 2 to 7 kPa.m^2/g) are in the ranges of straw pulp (respectively 6 to 8 km and 3 to 5 kPa.m^2/g) and bleached hardwood kraft pulp (respectively 7 to 10 km and 4 to 7 kPa.m^2/g). Tensile and burst strength of the hemp woody core pulps develop very rapidly with beating. The tear strength of test paper made from hemp woody core pulp is approximately 3.5 mN.m^2/g (which is about 50 percent lower than found for hardwood). This low tear strength, expected for short-fibered material, diminishes very slightly with beating, indicating that delignified unbeaten hemp woody core fibers cohere adequately, which is similarly found for straw pulp (Clark, 1985), with tear values in the same range. Not much effort is needed (either chemically or physically) to develop maximum attainable tear strength for hemp woody core sheets. The cohesion of unbeaten hardwood and softwood fibers is often too low to distribute the tear load effectively. Beating improves cohesion; tear strength of these pulps first increases, before it diminishes with further beating and some fiber shortening.

Brightness and opacity. The levels of brightness of unbleached and bleached hemp woody core pulp (respectively 40 and 75 percent) are in the range as reported for unbleached kraft and semi-bleached kraft hardwood pulps (Rydholm, 1985).

Typically, brightness and opacity are influenced by the removal of colored materials—the dominating source of discoloration being condensed and degraded lignin—and by beating, influencing the area of unbound fiber surfaces remaining available for light scattering, and resulting in a more transparent paper sheet (Singh, 1979; Rydholm, 1985).

Certainly, both brightness and opacity of the hemp woody core pulps diminish slightly with beating. However, relatively high lignin contents influence brightness more than PFI-beating does, while opacity of these pulps are in the same range. Bleaching drastically increases brightness, and decreases opacity (not only is some lignin removed, but also colored components are bleached), while PFI-beating lowers the opacity level slightly further.

Polymerization degree. In general, all strength properties of a pulp decrease with decreasing degree of polymerization (DP), the averaged number of glucose units per polymer (Rydholm, 1985). For hemp woody core pulp, its impact on paper strength, especially on tear factor, appears to be negligible when DP remains higher than 1,000 glucose units.

The DP first rises with decreasing pulp yield, which may be the effect of hemicellulose dissolution, resulting in an increased DP of the remaining carbohydrates. Thereafter, DP diminishes with decreasing pulp yield and more severe pulping conditions. The NaOH concentration is very important for cellulose depolymerization, as it is for cellulose degradation of hemp woody core, where a similar dependence was found (de Groot, van Dam, and van't Riet, 1995). This impact may be related to the relative low density of hemp woody core, facilitating cellulose accessibility and avoiding uneven NaOH concentrations (that may occur in more compact wood chips, and may explain lower impact of NaOH concentration on cellulose degradation and depolymerization in comparison to hemp woody core chips).

The DP diminishes faster than the cellulose content. At 170° C, the calculated velocity constant for depolymerization is about ten times the velocity constant found for cellulose degradation (de Groot, van Dam, and van't Riet, 1995). This is conceivable, as depolymerization proceeds mainly by alkaline hydrolysis, each fragmentation removing multiple glycosidic units from a cellulose molecule, while cellulose yield loss is mainly the result of peeling, degrading reducing-end units one by one and thus with a lower velocity rate. We calculated a similar difference between depolymerization and cellulose degradation rate, which corresponds to the data of Lai and Sarkanen (1967).

Conclusion

Hemp woody core pulp is comparable to wood pulp in many aspects. It is chemically and botanically related to hardwood fibers and not to straw fibers (de Groot et al., 1994). However, in beating response, hemp woody core pulp resembles the easily developing straw pulps, while tear factor and tensile strength are also in the same range. Burst factors measured for hemp woody core are in the range of the values found for bleached hardwood kraft pulp. Therefore, it is conceivable to further develop alkaline woody core pulping for similar purposes such as hardwood and straw pulp and as a component in pulp mixes for printing grade paper. This has to be verified and quantified (using varying percentages of hemp woody core pulp) in larger scale experiments.

RECOMMENDATIONS

Demonstration plant for hemp and other paper fiber sources in Europe

A pulping extruder is an effective and versatile device for high-yield pulping, using hemp bast fibers and comparable vegetable fibers. Extrusion pulping cuts the long fibers sufficiently, preventing spinning and knotting problems in modern stock preparation and paper machines. Unbleached and bleached hemp bast pulps were produced at a high-yield (80 percent) and at energy requirements as low as half of that of typical mechanical softwood pulp. The pulps produced on a laboratory scale display satisfactory qualities. The results have led to application research and feasibility projects for using hemp bast pulps for reinforcement of testliners and other disposable packaging materials, replacement of synthetic fibers in nonwovens, replacement of NBSK and BCTMP in printing and writing grade papers, and replacement of glass fiber in composite materials.

The practice of using a set of extruders and refiners in high-yield pulping makes the mill flexible and lowers the risks of failure compared to a single unit used for chemical pulping.

The trend of upgrading shorter fibers is likely to continue, implying that in the future more use will be made of high-yield hardwood (HYH) pulp, replacing chemical pulp in printing and writing grades, pulp qualities being almost identical. Hemp woody core could be developed as HYH pulp, with qualities that are equally promising as that of the test papers that were produced with alkaline hemp woody core pulp, both bleached and unbleached (good smoothness and strength characteristics).

Although the market for HYH pulp is growing, and hemp bast fibers can be processed efficiently with high-yield methods, high-yield pulping should be further developed for fiber hemp. Furthermore, using 75 to 85 percent of the hemp stem for paper, compared with 40 to 50 percent when chemical processes are employed, seems a better exploitation for any fibrous raw material. Also, other fibrous material, such as poplar, flax, kenaf, and straw could be pulped with little adaptation of the equipment.

Pilot plant studies for chemi-thermomechanical pulping can be started with one process unit of 10,000 ton pulp per year for extrusion pulping (bast fibers) and 25,000 ton pulp per year for chemi-thermomechanical pulping, using refiners (woody core fibers), which can be built out into a full-size mill of several units.

It would be more economical if the pulp mill could be combined with a paper mill. This enables efficient energy, water, and effluent management (using process heat for paper drying, sharing water, and fiber recovery systems).

We recommend starting with a demonstration plant of limited scale for these pilot plant studies. Although a demonstration plant could be placed anywhere as a nonintegrated unit, at this stage of research it might be more fruitful to build the plant close to an existing paper mill, to employ the papermaker's expertise.

REFERENCES

Ahlgren, L. (1991). "Kraft pulp: Cleaner still by year 2000." *Pulp and Paper International*, April 1991:59-61.
Anonymous. (1996). "Car-maker turns to cannabis—for fiber." *Nature*, 384:95.
Anonymous. (1997). "Mercer International Inc. plans to build Germany's first modern kraft pulp mill." *Tappi Journal*, 80(1):16.

Atack, D. (1978). "Advances in beating and refining." In *Fiber-Water Interactions in Papermaking Transactions of the Oxford Symposium,* pp. 261-295. London, UK: British Paper and Board Industry Federation.

Atchison, J. E. (1996). "Twenty-five years of global progress in nonwood plant fiber repulping." *Tappi Journal,* 79(10):87-95.

Baker, C. F. (1995). "Good practice for refining the types of fiber found in modern paper furnishes." *Tappi Journal,* 78(2):147-153.

Bohn, W. L. and Sferrazza, M. J. (1989). "Alkaline peroxide mechanical pulping: A revolution in high-yield pulping." In *Proceedings of the 1989 International Mechanical Pulping Conference,* pp. 184-200. CPPA, Montreal, Canada.

Bosia, A. (1975). "Hemp for refiner mechanical pulp." *Paper, World Research, and Development,* 1975:37, 41.

Bosia, A. and Nisi, D. (1978). "Utilizzazione integrale della canapa con i processi alcali-ossigeno e chemi-mechanico" (Italian). *Cellulosa e Carta,* 29(2):7-15.

Broderick, G., Paris, J., Valade, J. L., and Wood, J. (1996). "Linking the fiber characteristics and handsheet properties of high-yield pulp." *Tappi Journal,* 79(1): 161-169.

Clark, J. d'A. (1985). *Pulp technology and treatment for paper,* Second edition. San Francisco, CA: Miller Freeman Publications.

Cort, C. J. and Bohn, W. L. (1991). "Alkaline peroxide mechanical pulping of hardwoods." *Tappi Journal,* 74(6):79-84.

Dasgupta, S. (1994). "Mechanism of paper tensile-strength development due to pulp beating." *Tappi Journal,* 77(6):158-166.

de Groot, B. (1995). "Hemp pulp and paper production: Paper from hemp woody core." *Journal of the International Hemp Association,* 2(1):31-34.

de Groot, B. and Harsveld van der Veen, J. E. (1988). "Further research on the processing of hemp into paper" (Dutch). Wageningen Agricultural University, Netherlands, p. 67.

de Groot, B., van Zuilichem, D. J., and van der Zwan, R. P. (1988). "The use of nonwood fibers in the Netherlands." In *Proceedings of the 1988 International Nonwood Fiber Pulping and Papermaking Conference,* pp. 216-222. Beijing, China.

de Groot, B., van Dam, J. E. G., van der Zwan, R. P., and van't Riet, K. (1994). "Simplified kinetic modelling of alkaline delignification of hemp woody core." *Holzforschung,* 48(3):207-214.

de Groot, B., van Dam, J. E. G., and van't Riet, K. (1995). "Alkaline pulping of hemp woody core: Kinetic modelling of lignin, xylan, and cellulose degradation." *Holzforschung,* 49(4):332-342.

de Groot, B., van der Kolk, J. C., van Dam, J. E. G., and van't Riet, K. (1996). "Development of alkaline hemp woody core pulp and paper with beating, pulp yield, and composition." Submitted for publication.

de Groot, B., van der Kolk, J. C., van der Meer, P., van Dam, J. E. G., and van't Riet, K. (1997). "Alkaline swelling of hemp woody core chips." *Journal of Wood Chemistry and Technology,* 17(1/2):187-208.

Dewey, L. H. and Merrill, J. L. (1916). "Hemp hurds as papermaking material." *USDA Bulletin 404*, Washington, DC, 25 pp.

Dolk, M., Woemer, D., Lai, D., Kondo, R., and McCarthy, J. L. (1983). "Preliminary results of a study of the delignification of Western Hemlock using a 'flowthrough' reactor." In *Proceedings of the International Symposium on Wood and Pulping Chemistry*, pp. 146-149. Tsukuba Science City, Japan.

FAO. (1991). *The outlook for pulp and paper to 1995. Paper products, an industrial update.* Rome: Food and Agricultural Organization of the United Nations.

Ford, M. J. and Sharman, P. M. (1996). "HYH spells a good deal for the future." *Pulp and Paper International*, October 1996:29-39.

Iyengar, R. S. (1982). "Experiences in refining bagasse and other nonwood fibers." In *Nonwood Plant Fiber Pulping Progress Report 12*, pp. 25-30. Atlanta, GA: Tappi Press.

Janson, J. (1980). "Pulping processses based on autocausticizable borate." *Svensk Papperstidning*, 83(14):392-395.

Janson, J. (1992). "Fosfatkok—En ny och enkel process för massa ur gräs." Lecture (Swedish), *The Finnish Pulp and Paper Research Institute*, p. 8. Helsinki, Finland.

Janson, J., Jousimaa, T., Hupa, M., and Backman, R. (1994). "The use of *Festuca arundinaceae*: Pulping, bleaching, papermaking, and spent liquor recovery." In *The Seventh International Symposium on Wood and Pulping Chemistry, Proceedings*, Volume 1, pp. 278-282. Beijing, China.

Jeffers, L. and Malito, M. (1996). "On-line measurement of lignin in wood pulp by color shift of fluorescence, U.S. Patent 5,486,915. *Tappi Journal*, 79(4):44.

Jiménez, L., López, F., and Martinez, C. (1993). "Paper from sorghum stalks." *Holzforschung*, 47(6):529-533.

Kerekes, R. J. and Schell, C. J. (1995). "Effects of fiber length and coarseness on pulp flocculation." *Tappi Journal*, 78(2):133-139.

Lai, Y.-Z. and Sarkanen, K. V. (1967). "Kinetics of alkaline hydrolysis of glycosidic bonds in cotton cellulose." *Cellulose Chemistry and Technology*, 1:517-527.

Lecsek, P. (1994). "Pulp producers clean up in a big way with small boilers." *Pulp and Paper International*, December 1994:26-27.

Matussek, H., Pappens, R. A., and Cenny, J. (1996). "Unlucky 13th year for world production growth?" *Pulp and Paper International*, July 1996:22-25.

Meadows, D. G. (1995). "The pulp mill of the future: 2005 and beyond." *Tappi Journal*, 8(10):55-60.

Meadows, D. G. (1996). "Meadow Lake marks fourth year of zero liquid effluent pulping." *Tappi Journal*, 79(1):63-68.

Nordkvist, E., Graham, H., and Aman, P. (1989). "Soluble lignin complexes isolated from wheat straw *(Triticum arvense)* and red clover *(Trifolium pratense)*." *Journal of the Science of Food and Agriculture*, 48(3):311-321.

Page, D. H. (1994). "A note on the mechanism of tearing strength." *Tappi Journal*, 77(3):201-203.

Paulonis, M. A. and Krishnagopalan, A. (1988). "Kappa number and overall yield calculation based on digester liquor analysis." *Tappi Journal*, 71(11):185-187.

Pulliam, T. L. (1995). "Mills draw from growing number of non-chlorine, TEF options." *Pulp and Paper*, 69(9):75-83.

Rydholm, S. A. (1985) (reprint ed.). *Pulping processes*. Malabar, FL: R. E. Krieger Publishing Company.

Saltin, J. F. and Strand, W. C. (1992). "Improving the reliability of newsprint quality data using integrated factor networks for paper machine control and analysis." *Tappi Journal*, 75(12):55-61.

Sameshima, K. and Ohtani, Y. (1995). "What should be done between forestry and agriculture? A need for research and development on kenaf as a new model" In Gurnmerus Kirjapaino Oy (Ed.), *The Eighth International Symposium on Wood and Pulping Chemistry, Proceedings*, Volume III, pp. 273-276. Jyväskylä, Finland.

Sarkanen, K. V., Chang, H.-M., and Allan, G. G. (1967a). "Species variation in lignins. II. Conifer lignins." *Tappi Journal*, 50(12):583-587.

Sarkanen, K. V., Chang, H.-M., and Allan, G. G. (1967b). "Species variation in lignins. III. Hardwood lignins." *Tappi Journal*, 50(12):587-590.

Seisto, A. and Poppius-Levlin, K. (1995). "MILOX pulping of agricultural plants." In Gurnmerus Kirjapaino Oy (Ed.), *The Eighth International Symposium on Wood and Pulping Chemistry, Proceedings*, Volume I, pp. 407-414. Jyväskylä, Finland.

Seth, R. S. (1995). "The effect of fiber length and coarseness on the tensile strength of wet webs: A statistical geometry explanation." *Tappi Journal*, 78(3):99-102.

Seth, R. S. and Page, D. H. (1988). "Fibre properties and tearing resistance." *Tappi Journal*, 71(2):103-107.

Singh, R. P. (Ed.). (1979). *The bleaching of pulp*, Third revised edition. Atlanta, GA: Tappi Press.

Stefan, V. (1995). "Market pulp producers ride wave of worldwide industry recovery." *Pulp and Paper*, 69(13):87-93.

Strand, B. C., Saltin, J. L. F., and Bardwell, B. B. (1994). "Real-time quality analysis of newsprint paper machines." *Tappi Journal*, 77(4):145-149.

Stratton, R. A. (1991). "Characterization of fiber: Fiber bond strength from paper mechanical properties." In *Proceedings 1991 International Paper Physics Conference*, pp. 561-577. Atlanta, GA: Tappi Press.

Tikka, P. O. and Virkola, N.-E. (1986). "A new kraft pulping analyzer for monitoring organic and inorganic substances." *Tappi Journal*, 69(6):66-71.

Trinh, D. T. (1988). "The measurement of lignin in kraft pulping liquors using an automatic colorimetric method." *Journal of Pulp and Paper Science*, 14(1): J19-J22.

van den Ent, E. J. and Harsveld van der Veen, J. E. (1994). "Selective pulping of nonwood plant species: Pectin in and between fiber cell walls in plant tissue." In *1994 Tappi Pulping Conference*, pp. 749-757. Atlanta, GA: Tappi Press.

van Kemenade, M. J. J. M., van Roekel, G. J., van Berlo, J. M., Lips, S. J., de Groot, B., Zomers, F. H. A., de Vries, F. P., and Harsveld van der Veen, J. E. (1993). "Starting points and options for a Dutch demonstration pulp mill." In J. M. van Berlo (Ed.), *Paper out of hemp from Dutch soil* (Dutch), pp. 184-207. Wageningen, Netherlands, ATO-DLO.

van Roekel, G. J., Jr. (1995). "Chemimechanical pulping of fiber hemp for production of hemp-based printing and writing grade paper." In *Proceedings of the Symposium Bioresource Hemp,* Second Edition, pp. 442-455. Cologne, Germany: Nova-Institut.

van Roekel, G. J., Jr. (1996). "Bulk papermaking applications for bast fiber crops using extrusion pulping." In *Uses for non-wood fibers*, p. 9. Peterborough, UK: PIRA International.

van Roekel, G. J., Jr., Lips, S. J. J., Op den Kamp, R. G. M., and Baron, G. (1995). "Extrusion pulping of true hemp bast fiber (*Cannabis sativa* L.)." In *1995 Tappi Pulping Conference*, p. 9. Chicago, IL: Tappi Press.

Vanchinathan, S. and Krishnagopalan, G. A. (1995). "Kraft delignification kinetics based on liquor analysis." *Tappi Journal*, 78(3):127-132.

Vanchinathan, S. and Krishnagopalan, G. A. (1997). "Dynamic modeling of kraft pulping of southern pine based on on-line liquor analysis." *Tappi Journal*, 80(3): 123-133.

Wedekind, E. (1938). Neuere Methoden zur Verarbeitung von Laubhölzern und ähnlichen pflanzlichen Produkten auf Cellulose. *Wochenblatt für Papierfabrikation Sondernummer,* 1938:27-29.

Wedekind, E., Grasshof, H., and Müller, O. (1937). "Ueber den alkalischen Aufschluß von Laubhölzern und Holzaustauschstoffen" (German). *Der Papier-fabrikant Heft*, 51/52:513-516.

Wong, A. (1994). "Impact of biomass potassium on the operation of effluent-free agri-pulp mills." In *1994 Tappi Pulping Conference, Proceedings*, pp. 745-747. Atlanta, GA: Tappi Press.

Chapter 11

Hemp Seed: A Valuable Food Source

David W. Pate

INTRODUCTION

Cannabis is probably one of the first plants to have been used (and later cultivated) by people (Schultes, 1973). Throughout history and in separate parts of the world, hemp has often been an important plant, revered for its psychoactivity and useful for medicine, as a source of fiber, and for the food provided by its seed. The seed oil is particularly nutritious and its properties and potentials are herein explored.

The fruit of hemp is not a true seed, but an "achene," a tiny nut covered by a hard shell (Small, 1979; Paris and Nahas, 1984). These are consumed whole, used in food and folk medicinal preparations (Jones, 1995), or employed as a feed for birds and fishes. Whole hemp seed contains approximately 20 to 25 percent protein, 20 to 30 percent carbohydrates and 10 to 15 percent insoluble fiber (Theimer and Mölleken, 1995; Theimer, 1996), as well as a rich array of minerals, particularly phosphorous, potassium, magnesium, sulfur, and calcium, along with modest amounts of iron and zinc (Jones, 1995; Wirtschafter, 1995), the latter of which is an important enzyme cofactor for human fatty acid metabolism (Erasmus, 1993). It is also a fair source of carotene, a "Vitamin A" precursor, and is a potentially important contributor of dietary fiber. Most hemp seed also contains approximately 25 to 35 percent oil, although one variety grown in Russia, known as "olifera," reportedly contains 40 percent (Small,

---

Thanks to my collaborators J. C. Callaway and J.-L. Deferne, as co-authors of articles upon which this work is based.

1979; Mathieu, 1980) and a Chinese variety was claimed to slightly exceed this figure (Jones, 1995).

This highly polyunsaturated oil has uses similar to that of linseed oil (e.g., fuel for lighting, printer's ink, wood preservative), but also has been employed as a raw material for soaps and detergents (Olschewski, 1995) and as an emollient in body care products (Rausch, 1995). However, it is the nutritional qualities of the oil that are particularly important. The crushed seed by-product is suitable for animal feed as well as a human staple (Grinspoon and Bakalar, 1993; Small, 1979; Paris and Nahas, 1984), due to its spectrum of amino acids (Odani and Odani, 1998), including all eight of those essential to the human diet (Jones, 1995; Wirtshafter, 1995), as well as carbohydrates and a small amount of residual oil. Its protein is primarily edestin (St. Angelo, Yatsu, and Altschul, 1968), a highly assimilable globular protein of a type similar to the albumin found in egg whites and blood. However, heat-treating whole hemp seed denatures this protein (Stockwell, Dechary, and Altschul, 1964) and renders it insoluble, possibly affecting digestibility.

An ideal seed hemp variety would produce a high yield of seed (normally 0.5 to 1.0 t/ha) containing a high percentage of good quality oil. Highly branched varieties are usually preferred. For seed production, male plants are sometimes removed after pollination has occurred, in order to leave more space for female plants. Mathieu (1980) has noted that seed yield can be doubled using monoecious varieties, although this sexual type suffers some inbreeding depression. Cultivation of these strains reportedly has produced up to approximately 1.5 metric tons of seed per hectare, but lower yields are generally expected (Mathieu, 1980; Höppner and Menge-Hartmann, 1994). Highest seed yields are reportedly obtainable with unisex female varieties such as Uniko-B (Bócsa, 1995), although FIN-314, a recently developed variety that is nonbranching and dioecious, has been demonstrated to produce record yields of up to two tons per hectare in Finland (Callaway, 1998). The number of flowers per plant and, therefore, the quantity of seed produced, can be increased by "topping" the plants when they are 30 to 50 cm high. Maximum seed yield requires that hemp be sown at a much lower density than for fiber (Reichert, 1994). However, weeds can prosper if planting density is too sparse (e.g., $25/m^2$).

EXTRACTION METHODS

Extraction of oil from hemp seed is not being carried out on a large scale at the present time. That which is being processed is sometimes relatively unhomogenous, having mature seeds mixed with green ones. This is due to the difficulty of finding the optimal time for harvesting, since not all seeds reach maturity simultaneously, especially in hemp undeveloped for seed production. The presence of unripe seeds not only increases seed crop moisture content, it also lowers oil yield and modifies its taste.

After harvest, hemp seed undergoes a drying process that reduces its moisture content to 10 percent or less, so as to prevent sprouting during storage. Batches of this material are then fed into a hydraulic screw press and a pressure of 500 bars is progressively applied, resulting in only a minor elevation (to 50° C) in temperature. Best quality oil is obtained from the first fractions recovered. Approximately 35 percent of the available oil remains in the seed cake (Jones, 1995). The pressing process is sometimes repeated with this crushed residue to obtain a small additional amount of oil, although quality is decreased. This "cold pressing" does not allow an extraction yield equal to that of techniques employing high temperatures, but it has the advantage of minimizing degradative changes in the oil. A small amount of oil is also unrecovered during the subsequent filtration process.

Solvents are sometimes used to extract the oil, but this causes contamination with undesirable residues. An approach that avoids this drawback is the use of supercritical fluids or liquified gases that rapidly evaporate completely at atmospheric pressure. Carbon dioxide is the leading candidate for this method as it is relatively cheap and inherently nontoxic. An additional advantage to this method is that the resulting seed residue is completely devoid of residual oil, allowing its protein content to be processed into an easily storable dry product similar, but superior, to the bulk texturized vegetable protein currently manufactured from soy. Further oil refining procedures should be avoided in order to preserve the native qualities of this product. Bottling must occur quickly, and filling under nitrogen into opaque bottles, then refrigerating, offers significant protection against oil degradation due to oxidation and the action of light, although freezing is necessary for long-term storage. Addition of anti-oxidants extends shelf life of the product at room temperature (McEvoy, Edwards, and Snowden, 1996).

OIL COMPOSITION AND PROPERTIES

Nonrefined hemp seed oil extracted by cold-pressing methods varies from off-yellow to dark green and has a pleasant nutty taste, sometimes accompanied by a touch of bitterness. The seed (and therefore the extracted oil) normally does not contain significant amounts of psychoactive substances (Paris and Nahas, 1984; Vieira, Abreu, and Valle, 1967). Trace amounts of THC, sometimes found upon analysis, are probably due to contamination of the seed by adherent resin or other plant residues (Matsunaga et al., 1990; Mathé and Bócsa, 1995), although reports to the contrary exist (e.g., Patwardhan, Pundlik, and Meghal, 1978). However, these traces have been shown to produce positive urinalysis results, particularly upon regular use of the oil (Callaway et al., 1997).

Analytical data reported for the fatty acid composition of hemp seed oil (Weil, 1993; Kralovansky and Marthné-Schill, 1994; Höppner and Menge-Hartmann, 1994; Rumyantseva and Lemeshev, 1994; Theimer and Mölleken, 1995; Wirtshafter, 1995; Callaway and Laakkonen, 1996; Callaway, Tennilä, and Pate, 1996; Mölleken and Theimer, 1997) reveals that it is unusually high in polyunsaturated fatty acids (up to 80 percent), while its content in saturated fatty acids (approximately 10 percent) compares favorably with the least saturated commonly consumed vegetable oils (see Table 11.1). This high degree of unsaturation explains its sensitivity to oxidative rancidity, as the chemical "double-bonds" that provide such unsaturation are vulnerable to reaction with atmospheric oxygen. This degradation is accelerated by heat or light. For this reason, and to prevent the formation of *trans*-fatty acids (Wolff, 1993), the oil is unsatisfactory for prolonged or high temperature ($\geq 180°$ C) frying, although moderate heat for short periods is probably tolerable. It is best consumed as a table oil, on salads or as a butter/margarine substitute for dipping bread, similar in use to olive oil. Proper steam sterilization of the seed probably does not cause significant damage to the oil, but does destroy the integrity of the seed, allowing penetration by air and molds. If this procedure is legally required, it should be done at a bonded facility immediately before release of the seed for further processing. By the same reasoning, one should avoid eating whole hemp seed that has been subjected to any cooking process, unless reasonably fresh.

Table 11.1. Profile of hemp seed compared to common edible oils (% total fatty acids).

Less healthy/Chemically stable <---> More nutritious/Chemically unstable					
	"Saturated"		"Monounsaturated"	"Polyunsaturated"	
	Palmitic (C16:0)	Stearic (C18:0)	Oleic (C18:1ω9)	Linoleic (C18:2ω6)	Linolenic (C18:3ω3)
Hemp	6-9	2-3	10-16	50-70	15-25
Soy	9	6	26	50	7
Canola	0	7	54	30	7
Wheatgerm	0	18	25	50	5
Safflower	0	12	13	75	0
Sunflower	0	12	23	65	0
Corn	0	17	24	59	0
Cottonseed	0	25	21	50	0
Sesame	0	13	42	45	0
Peanut	0	18	47	29	0
Avocado	0	20	70	10	0
Olive	0	16	76	8	0
Palm	85	0	13	2	0
Coconut	91	0	6	3	0

Source: Adapted from Erasmus, 1993.

Two polyunsaturated essential fatty acids (EFAs), linoleic acid (C18:2ω6) or "LA" and linolenic acid (C18:3ω3) or "LNA," cannot be manufactured by the human body and must come from dietary sources (Erasmus, 1993). They account for approximately 50 to 70 percent and 15 to 25 percent respectively, of the total seed fatty acid content (Deferne and Pate, 1996). Such a 3:1 balance has been claimed optimal for human nutrition (Erasmus, 1993) and is apparently unique among the common plant oils (see Table 11.1). *Cannabis* seed from tropical environments seems to lack significant quantities of LNA (Ross et al., 1996; Theimer and Mölleken, 1995) and there seems a general bias evident toward a more unsaturated fatty acid content among high-latitude origin *Cannabis* seed specimens. This may reflect a regional evolutionary selection pressure. The possible influence of local environmental inputs is presently unknown, but since latitude of cultivation is known to influence degree of fatty acid unsaturation in uniform genetic strains of other oilseed plants (de Meijer, 1996), further experiments are necessary to differentiate these two influences. In either case, the metabolic desaturation of plant oils is energetically expensive and occurs toward the finish of fatty acid formation (i.e., is not necessary to achieve carbon-chain lengthening). For this reason, and since this type of oil remains more mobile at the relatively lower winter temperatures of the Nordic climate (also storing more energy per molecule), it would seem that some local advantage (e.g., survival in extreme cold or allowance of earlier germination) might be attributable to its presence.

The range of results found in some analyses may be attributable to differences in crop ripeness, since formation of polyunsaturated fatty acids is incomplete in immature *Cannabis* seed (Ross et al., 1996). This suggests that a maximum ripening of the seed and the culling of immature seed are important considerations for the production of a quality oil. Likewise, proper seed sampling criteria are also crucial for representative analyses.

CRITICAL ENZYME

In contrast to the shorter-chain and more saturated fatty acids, EFAs serve not as energy sources, but as raw materials for human cell structure and as precursors for biosynthesis of many of the body's regulatory biochemicals (Spielmann et al., 1988). Products of these syntheses include the powerful, short-lived, hormone-like prostaglandins and, coincidentally, the recently discovered THC-receptor ligand known as "anandamide" (Hansen, 1994). "Series 1" and "Series 2" prostaglandins are produced from LA (Erasmus, 1993) via its conversion by the enzyme *delta*-6-desaturase (see Figure 11. 1) to *gamma*-linolenic acid (GLA, $18:3\omega6$). Similarly, "Series 3" prostaglandins are produced from LNA via its preliminary conversion by this same enzyme to stearidonic acid (SDA, $18:4\omega3$). These three series fulfill numerous vital biochemical roles, ranging from control of inflammation processes and vascular tone to initiation of contractions during childbirth.

Figure 11.1. Metabolic pathways for the production of prostaglandins from fatty acids. Not known are additional metabolic steps after the production of GLA and SDA.

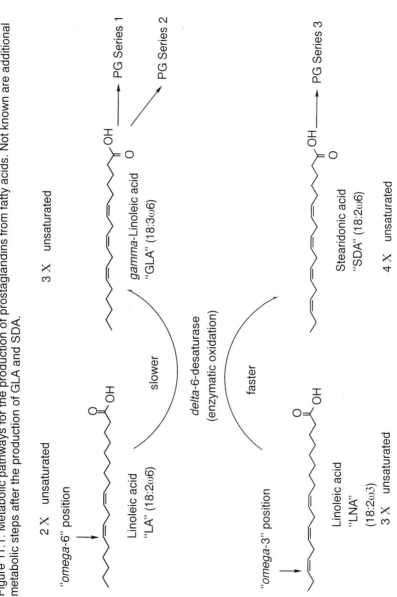

Source: Callaway, Tennila, and Pate (1996).

The metabolic conversion of LA to GLA is slow in mammals. In addition, this enzymatic activity can be weak or lacking due to hereditary defects or can be inhibited by alcoholism, physiological stress, and some degenerative diseases (e.g., hypertension, diabetes, etc.), especially in the elderly. If impairment of this enzyme is the problem, dietary supplementation with GLA (Horrobin, 1990a, 1990b) and in more severe cases, SDA can compensate for this deficiency. In addition, the surfeit of LA (combined with the low LNA levels) usually found in those with the usual health-conscious "polyunsaturated" diet may impair *delta*-6-desaturase conversion of both LA and the little LNA present (Spielmann et al., 1988). Unfortunately, very few foods contain GLA or SDA, and available concentrated supplements of these fatty acids are expensive.

GLA IMPORTANCE

GLA is found in minute quantities in most fats of animal origin (Horrobin, 1990a, 1990b). Oats and barley also contain small amounts. Human milk contains some GLA (Carter, 1988), but any significance is probably overshadowed (Erasmus, 1993) by the greater presence of its metabolic derivative dihomo-*gamma*-linolenic acid or "DGLA" ($C20:3\omega6$).

The potential physiological effects of GLA supplements have been extensively investigated only recently. Its alleviating action on psoriasis, atopic eczema, and mastalgia are already well documented and GLA preparations are now frequently prescribed for the treatment of the latter two disorders. GLA has also been under investigation for its beneficial effects in psychiatric, cardiovascular, and immunological disorders (Horrobin 1990a, 1990b, 1992). However, daily excess consumption of GLA, especially in those not needing this supplement, may allow the accumulation of a metabolic excess of arachidonic acid and, thereby, possibly promote inflammation, thrombosis, or immunosuppression (Phinney, 1994).

GLA is available exclusively in health food shops or pharmacies, mostly as soft gelatin capsules, and is not found in oils usually consumed by most people. Good sources of GLA include the blue-green alga *Spirulina* (~ 1 percent of dry weight) and (see Table 11.2) evening primrose (*Oenothera biennis* L.) oil, black currant seed oil

Table 11.2. Oil profiles of major GLA sources (% total fatty acids).

	Palmitic (C16:0)	Stearic (C18:0)	Oleic (C18:1ω9)	Linoleic (C18:2ω6)	Linolenic (C18:3ω3)	gamma-Linolenic (C18:3ω6)
Hemp	6-9	2-3	10-16	50-70	15-25	1-6
Evening Primrose	4-12	1-7.5	4-12	65-72	0	3-15
Black Currant	6-7	1-2	9-11	45-60	12-15	15-19
Borage	~11	~4	~16.5	~37	<1	~23
Fungus (Mucor)	9-12	1-2	20-40	18-20	0	20-40

Source: Deferne and Pate (1996).

(*Ribes nigrum* L.), borage (*Borago officinalis* L.) oil, and some fungal (*Mucor*) oils. Hemp seed oil from sterilized seed analyzed in the United States contained 1.7 percent GLA (Weil, 1993; Wirtshafter, 1995). Callaway and Laakkonen (1996) reported substantial amounts of GLA (4 percent) in seed of their early-blooming "high-latitude hybrid" *Cannabis* (FIN-314). Higher levels (5.69 percent) have been measured by this group (Callaway, Tennilä, and Pate, 1996) and confirmed (up to 6 percent) by German investigators (Theimer, 1996), although it is apparently rare in most tropical varieties of *Cannabis* (ElSohly, 1996). However, absolute amounts of GLA are not the only criteria for ranking the desirability of an oil. The particular arrangement of these fatty acids on glycerol (as the natural triglyceride), as well as differences in possible toxicity among the various oils, may be important considerations (Horrobin, 1994).

SDA SUPPORTING ROLE

SDA may also function as an important human dietary supplement, but only for people with rather severe deficits in their *delta*-6-desaturase function. This is because SDA formation in the human body is faster than that of GLA (assuming adequate LNA levels), so not much of the former supplement is needed until SDA/GLA processing becomes quite inhibited.

However, relatively few people suffer from such a defect in this enzyme compared to the nearly universal lack of adequate LNA levels in the diet. For those whose levels of SDA are low due only to diet, supplementation with LNA is sufficient to restore the necessary balance. A severe LNA deficit is best acutely treated with flax (*Linum usitatissimum* L.) seed (fresh linseed) oil, although it is unsuitable for prolonged consumption due to an imbalance in its LA (e.g., 14 percent) to LNA (e.g., 58 percent) content, a ratio approximately equal, but inverse, to that of hemp seed (Erasmus, 1993).

SDA seems to have a very limited presence in domesticated plants. Black currant seed oil reportedly contains up to 9 percent SDA, although 2 to 4 percent is more usual (Clough, 1996). However, Callaway, Tennilä, and Pate (1996) reported significant amounts (up to 2 percent) of SDA in hemp seed, a compound heretofore unknown from this genus.

FUTURE PROSPECTS

No other single plant source offers a more favorable human dietary balance of the two essential fatty acids, combined with an easily digestible complete protein. Even though hemp seed oil contains only relatively small amounts of GLA/SDA when compared to more established sources, it is probably sufficient for many of those who cannot efficiently convert EFAs. This also helps EFA consumers not needing such supplements to avoid excessive daily GLA/SDA intake. In addition, because of its ease of cultivation, *Cannabis* may possess the potential to become an alternative raw material source for the economical production of isolated forms of GLA/SDA. The substantial presence of both GLA and SDA in the seed oil of FIN-314 (Callaway, Tennilä, and Pate, 1996), combined with its high oil output (37 percent), suggests this variety as a candidate for such an application.

Questions arise concerning the reasons which have so far prevented a more extensive consumption of hemp seed products. It is possible that the historically significant uses of hemp (i.e., fiber, medicine, whole seed, psychoactive drug) took priority over its potential utilization as a source of isolated oil and protein. Second, many other plant sources have been found more adequate in terms of yield and chemical stability of their oil. Likewise, the protein content of pressed hemp seed contains residual labile oil residues that render the product unfit for long-term storage. Finally, the relatively recent "anti-drug" ban on hemp cultivation in many countries has prevented food scientists from investigating its little-known nutritional value in more depth, and discovering the wide range of potential uses for this seed.

Much work remains to be undertaken with the existing cultivars, as well as indigenous landraces and feral strains. A major research priority must be the full characterization of oils obtained from diverse hemp sources. There exists considerable potential for development of varieties providing larger yields of seed containing a higher oil content with a consistent fatty acid profile. Knowledge of environmental influences on seed quality and the development of improved agricultural methods will also contribute to the future success of this plant. In addition, important questions remain concerning this oil's physicochemical properties, triglyceride structures, and physiological effects, as well as the methods of extraction and storage that are most economical and best suited to preserve its unique nutritional qualities.

REFERENCES

Bócsa, I. (1995). Personal communication.
Callaway, J. C. (1998). Personal communication.
Callaway, J. C. and Laakkonen, T. T. (1996). "Cultivation of *Cannabis* oil seed varieties in Finland." *Journal of the International Hemp Association,* 3(1):32-34.
Callaway, J. C., Tennilä, T., and Pate, D. W. (1996). "Occurrence of 'omega-3' stearidonic acid (*cis*-6,9,12,15-octadecatetraenoic acid) in hemp (*Cannabis sativa* L.) seed." *Journal of the International Hemp Association,* 3(2):61-63.
Callaway, J. C., Weeks, R. A., Raymond, L. P., Walls, H. C., and Hearn, W. L. (1997). "A positive THC urinalysis from hemp *(Cannabis)* seed oil." *Journal of Analytical Toxicology,* 21(4):319-320.
Carter, J. P. (1988). "*Gamma*-linolenic acid as a nutrient." *Food Technology,* 42(6): 72-82.
Clough, P. (1996). Personal communication.
Deferne, J.-L. and Pate, D. W. (1996). "Hemp seed oil: A source of valuable essential fatty acids." *Journal of the International Hemp Association,* 3(1):1, 4-7.

ElSohly, M. (1996). Personal communication.
Erasmus, U. (1993). *Fats that heal, fats that kill.* Alive Books, 7436 Fraser Drive, Burnaby, BC, Canada VSJ5B9.
Grinspoon, L. and Bakalar, J. B. (1993). The history of *Cannabis.* In *Marihuana, the forbidden medicine,* pp. 1-23. New Haven, CT: Yale University Press.
Hansen, H. S. (1994). "New biological and clinical roles for the n-6 and n-3 fatty acids." *Nutrition Reviews,* 52:162-167.
Höppner, F. and Menge-Hartmann. (1994). "Anbauversuche zur Stickstoffdüngung und Bestandesdichte von Faserhanf." *Landbauforschung Völkenrode,* 44(4): 314-324.
Horrobin, D. F. (Ed.) (1990a). *Omega-6 essential fatty acids.* New York, NY: Wiley-Liss.
Horrobin, D. F. (1990b). "*Gamma*-linolenic acid: An intermediate in essential fatty acid metabolism with potential as an ethical pharmaceutical and as a food." *Reviews in Contemporary Pharmacotherapy,* 1:1-45.
Horrobin, D. F. (1992). "Nutritional and medical importance of *gamma*-linolenic acid." *Progress in Lipid Research,* 31(2):163-194.
Horrobin, D. F. (1994). "Natural does not equal safe." *Pharmaceutical Technology Europe,* December, pp. 14-15, 38.
Jones, K. (1995). *Nutritional and medicinal guide to hemp seed.* Rainforest Botanical Laboratory, P. O. Box 1793, Gibsons, BC, Canada VON1VO.
Kralovansky, U. P. and Marthné-Schill, J. (1994). "Data composition and use value of hemp seed" (Hungarian with English summary). *Novenytermeles,* 43(5):439-446.
Mathé, P. and Bócsa, I. (1995). "Can THC occur in hemp seed oil?" *Journal of the International Hemp Association,* 2(2):59.
Mathieu, J.-P. (1980). "Chanvre (hemp)." *Technique Agricole,* 5:1-10.
Matsunaga, T., Nagatomo, H., Yamamoto, I., and Yoshimura, H. (1990). "Identification and determination of cannabinoids in commercially available *Cannabis* seed." *Eisei Kagaku,* 36(6):545-547.
McEvoy, C., Edwards, M., and Snowden, M. (1996). "An overview of antioxidant, preservative and solvent excipients used in the pharmaceutical industry." *Pharmaceutical Technology Europe,* 8(6):36-40.
Meijer de, E. P. M. (1996). Personal communication.
Mölleken, H. and Theimer, R. R. (1997). "Survey of minor fatty acids in *Cannabis sativa* L. fruits of various origins." *Journal of the International Hemp Association,* 4(1):13-17.
Odani, S. and Odani, S. (1998). "Isolation and primary structure of a methionine- and cystine-rich seed protein of *Cannabis sativa.*" *Bioscience, Biotechnology, and Biochemistry,* 62(2):650-654.
Olschewski, M. (1995). "Umweltvertragliche Tenside fur Wasch-und Reingungsmittel auf Naturstoffbasis." In *Bioresource Hemp,* Second Edition, pp. 562-567. Nova-Institute, Rosenstr. 53,50678 Cologne, Germany.
Paris, M. and Nahas, G. G. (1984). "Botany: The unstabilized species." In *Marihuana in science and medicine,* pp. 3-36. G. G. Nahas (Ed.). New York: Raven Press.

Pathwardhan, G. M., Pundlik, M. D., and Meghal, S. K. (1978). "Gas-chromatographic detection of resins in *Cannabis* seeds." *Indian Journal of Pharmacy Science*, 40(5):166-167.
Phinney, S. (1994). "Potential risk of prolonged *gamma*-linolenic acid use." *Annals of Internal Medicine*, 120(8):692.
Rausch, P. (1995). "Verwendung von Hanfsamenöl in der Kosmetik." In *Bioresource Hemp*, Second Edition, pp. 556-561. Nova-Institute, Rosenstr. 53,50678 Cologne, Germany.
Reichert, G. (1994). "Government of Canada report on hemp." *Bi-weekly Bulletin*, 7(23):1-8.
Ross, S. A., ElSohly, H. N., ElKashoury, E. A., and ElSohly, M. A. (1996). "Fatty acids of *Cannabis* seeds." *Phytochemical Analysis*, 7:279-283.
Rumyantseva, L. G. and Lemeshev, N. K. (1994). "Current state of hemp breeding in the C.I.S." *Journal of the International Hemp Association*, 1(2):49-50.
Schultes, R. E. (1973). "Man and marijuana." *Natural History*, August, pp. 59-82.
Small, E. (1979). "Practical and natural taxonomy for *Cannabis*." Volume 1. In E. Small (Ed.). *The species problem in Cannabis*, pp. 171-211. Toronto, Canada: Corpus.
Spielmann, D., Bracco, U., Traitler, H., Crozier, G. Holman, R., Ward, M., and Cotter, R. (1988). "Alternative lipids to usual *omega*-6 PUFAs: *Gamma*-linolenic acid, *alpha*-linolenic acid, stearidonic acid, EPA, etc." *Journal of Parenteral and Enteral Nutrition*, 12(6):111S-123S.
St. Angelo, A. J., Yatsu, L. Y., and Altschul, A. M. (1968). "Isolation of edestin from aleurone grains in *Cannabis sativa*." *Archives of Biochemistry and Biophysics*, 124:199-205.
Stockwell, D. M., Dechary, J. M., and Altschul, A. M. (1964). "Chromatography of edestin at 50 degrees." *Biochimica Biophyset Actaica*, 82:221-230.
Theimer, R. R. (1996). Personal communication.
Theimer, R. R. and Mölleken, H. (1995). "Analysis of the oil from different hemp cultivars—perspectives for economical utilization." In *Bioresource Hemp*, Second Edition, pp. 536-543. Nova-Institute, Rosenstr. 53,50678 Cologne, Germany.
Vieira, J. E., Abreu, L. C., and Valle, J. R. (1967). "On the pharmacology of the hemp seed oil." *Medicina et Pharmacologia Experimentalis: (International Journal of Experimental Medicine)*, 16:219-224.
Weil, A. (1993). "Therapeutic hemp oil." *Natural Health*, March/April, pp. 10-12.
Wirtshafter, D. (1995). "Nutrition of hemp seeds and hemp seed oil." In *Bioresource Hemp*, Second Edition, pp. 546-555. Nova-Institute, Rosenstr. 53,50678 Cologne, Germany.
Wolff, R. L. (1993). "Heat-induced geometrical isomerization of α-linolenic acid: Effect of temperature and heating time on the appearance of individual isomers. *Journal of the American Oil Chemists Society*, 70(4):425-430.

Index

Page numbers followed by the letter "i" indicate illustrations; those followed by the letter "t" indicate tables.

Abamectin, biorational pesticide, 125
Abuscreen, 47
Accreted leaflets, inbreeding, 163
AccuPinch, 51
"Achene" nut, 243
Achenes, 135. *See also* Seeds
"Acid rain," 122
Acronicta americana, hemp pest, 113
Active charcoal, 189
Aculops cannabicola, hemp pest, 116
Adaption latitude (provenance), 137
Affinity, recombinant antibodies, 47
Afghan *Cannabis*, 14
Afghani #1, drug strain, 147
Afghanistan, *indica* landraces, 147
Aggregation pheromone, biorational pesticide, 126
Agrobacterium tumefaciens, hemp disease, 121
Agrochemical immunoassay, 44
Agromyra strigata, hemp pest, 114
Agromyza reptans, hemp pest, 114
Agronomic management, need for new, 61
Agrotis ipsilon, hemp pest, 115
Air pollution, hemp diseases, 122
Alarm pheromone, biorational pesticide, 126
Alfalfa mosaic virus (AMV), 114,121
Alkaline peroxide mechanical pulping (APMP), wood chips, 228
Alkanes, 22
American dagger (*Acronicta americana*), hemp pest, 113

Amides, synthetic chemical pesticides, 126
Amino acids, in whole hemp seed, 243
"Anandamide," THC-receptor ligand, 248
Anemophilous breeding, 3,15
Anthesis, 71
Anthracnose, cause of, 119
Antibiotic properties, cannabinoids, 30
"Anti-drug" ban effects on hemp cultivation, 253
Antifeedant system, 30
Antigen, 47
Antisera, 45
Ants, hemp pests, 112
Aphid midge (*Aphidoletes aphidimyza*), biocontrol predator, 123
Aphids, hemp pests, 114,116
  biocontrol of, 123
  drug biotype and, 110
Aphis fabae, hemp pest, 114
Apparatus, GC/MS method, 55
Arabis mosaic virus (ArMV), 121
Arctia caja, hemp pest, 113
Argentine sunflower virus, 115
Armyworms, hemp pests, 113
  NVP for, 123
Asparagus officinalis, 207
Assay formats, 47-57
Atractomorpha crenulata, hemp pest, 115
Autographa gamma, hemp pest, 113

257

Average radiation-use efficiency
(RUE), 79-80,88,93
definition of, 92
kenaf, 103

Bacillus thuringiensis (BT),
microbial pesticide, 123
Bacteria, hemp disease, 121
Bark
chemical content of, 72
kenaf, 104
separation from core, 78-79
stem dry matter, 98-99
Bast
chemical content, 72
early use of, 85
fiber content, 72-73
density and, 67
quality criterion, 76
"Beam" test, THC content, 175
Beet armyworm (Spodoptera
exigua), hemp pest, 113,115
Beet webworm (Loxostege
sticticalis), hemp pest, 113
Beetles, biocontrol predator, 123
Bemisia argentfolii, hemp pest, 114
Bemisia tabaci, hemp pest, 114
Beniko, hemp variety, 63,206
Bertha armyworm (Mamestra
configurata), hemp pest, 113
Beta counter, 48
Bhang aphid (Phorodon cannabis),
hemp pest, 114
Bialobrzeskie, hemp variety, 63,206
Biocides, 64,125
Biocontrol
disease and pest management,
122-124
predator, 122-123
Biogenesis cannabinoids, 22-24,
23i,32-34
Biogenesis Inc., clone THC-003,
45t,49,49i,50i
Biorational pesticides, 125

Bioreactors, 195,196
Birds, hemp pests, 110,116
BiVis (twin-screw) process, pulp
extrusion, 226
Black bean aphid (Aphis fabae),
hemp pest, 114
Black currant seed oil (Ribes
nigrum L.), 250,251t,252
Black mildew, cause of, 120
Bleached alkaline-peroxide
mechanical pulp (APXP),
227,228,229,230
Bleached chemi-thermomechanical
pulp (BCTMP), spruce, 229
Bleaching, ATO-DLO test, 228
Blight disease, causes of, 119,120,121
Boehringer Mannheim, Frontline, 51
Bolognese, hemp variety, 62
Bordeaux mixture, biorational
pesticide, 125
Botryosphaeria obtusa, stalk cankers,
118
Botrytis cinerea, 87-88,117,124
Bracts, 4,5,25,34
"Breaker," 75
Bredemann, G., 153,164
Bredemann method, 63,205
Brightness and opacity, woody core
pulp, 235-236
Broomrape (Orobanche ramosa),
root rot, 118,122,178
Brown blight, cause of, 119
"Brown fleck disease," 120
Brown leaf spot, cause of, 120
Bud worms, biocontrol of, 123
Bugs, hemp pests, 114-115
Burdock borer (Papaipema
cataphracta), 111
Byronia, inherited sex, 154,156,172

C_4-grass (Miscanthus sinensis), 103
Cabbage curculio (Ceutorhynchus
rapae), hemp pest, 112,113
Cabbage moth (Mamestra
brassicae), hemp pest, 113

Callus, hemp reproduction, 189-190, 191t,191i-194i,195
Calocoris norrvegicuss, hemp pest, 114-115
Cannabaceae family, 13
Cannabichromene (CBC), 24
 pigmentation, 32
Cannabidiol (CBD), 23
 diagnostic antibodies, 44,45t
 and UV-B, 31
Cannabigerol (CBG), 23
Cannabinoids
 early use of, 85
 in vitro synthesis of, 196-197
 traditional analysis of, 44
Cannabinol (CBN), diagnostic antibodies, 44,45t
Cannabis
 cell culture, secondary metabolites, 195
 DNA polymorphism in, 198
 gene pool, 62-63,134,138,140t
 history, 8-12
 subspecies (ssp.), 135-136
Cannabis afghanica, 109
Cannabis indica
 characterized, 13-14,109
 provenance, 139
 psychoactive strains, 136
 wide leaflets, 135
Cannabis ruderalis
 characterized, 14
 weedy form, 135
Cannabis sativa
 cell culture, 195,196-197
 characterized, 13
 early use of, 8,9i,10i,11i,12,85
 fiber/seed crops, 136
 genetic variation in, 206
 narrow leaflets, 135
 PRC analysis, 201
 provenance, 139
 RFLP analysis, 200-201
 sex determination, 204-205,208
Cannabivarina, cannabinoid type, 44

Canopy
 establishment, light and, 88,89,90,91t
 fungal disease and, 109-110
 plant density and, 68,88
 senescence, 80,88,89
 thermal time, 66
Carbamates, synthetic chemical pesticides, 126
Carmagnola, hemp variety, 57,62
 genetic variation in, 202,203i,206,207t
 Italian cultivar, 63,160,177
Casein hydrolisate, 196
Caterpillars, hemp pest, 110-111,113
Cell culture, 185,194
Cellulose, 72
 degradation, woody core pulp, 236
 hemp content, 72,87
Central Asia, hemp origin, 6
Ceutorhynchus pleurostigma, hemp pest, 112
Ceutorhynchus quadridens, hemp pest, 112
Ceutorhynchus rapae, hemp pest, 112,113
Ceutorhynchus roberti, hemp pest, 112
Ceutorhynchus suleicollis, hemp pest, 112
Chaetocnema concinna, hemp pest, 112
Chaetocnema hortensis, hemp pest, 112
Charcoal rot, cause of, 119
Chemical controls, disease process management, 125-126
Chemical pollution, papermaking industry, 86
Chemi-thermomechanical pulp (CTMP), spruce, 228
China
 early uses of *Cannabis*, 8
 growth of hemp, 86
Chinese fiber-type *Cannabis*, 12,62

Chinese hemp, 153,159,161,162,178
Chip size, ATO-DLO woody core test, 232
Chloealtis conspersa, hemp pest, 115
Chlorinated hydrocarbons, synthetic chemical pesticides, 126
Chrotogonus saussurei, hemp pest, 115
Chrysoperla carnea, biocontrol predator, 122-123
Cladosporium species, stalk cankers, 118
Climate, fiber hemp requirement, 69
Clonal propagation, 16,47,186-187, 189
Clone THC-003, evaluation of, 45t,49,49i,50i
Close stands, 2
Cockroaches, hemp pest, 115
"Cold pressing," hemp seed oil, 245
Color, hemp quality, 75
Common purple grackle, hemp pest, 116
Common stalk borer *(Papaipema nebris),* 111
Commonwealth Mycological Institute, 120
Compact inflorescence, gene for, 159
Competitive Enzyme-linked immunosorbent assay (ELISA), 48-51,53,56
Concentration limit, THC, 50,56,175
Confirmatory analysis, GC/MS method, 54,55
Coniothyrium cannabinum, stalk cankers, 118
Content, hemp trait, 64
Contrast kit, 51
Core
  chemical content of, 72,73
  separation from bark, 78-79
Corn borers *(Ostrinia nubilalis),* 110
Cossus cossus, 111
Cotton
  hemp replacement, 62,86
  industry, criticism of, 86

Cotton bellworm *(Heliothis armigera),* hemp pest, 113
Cottoncushion scale *(Icerya purchasi),* hemp pest, 114
Crickets, hemp pest, 115
Crop
  characteristics, plant density impact, 69,98,100i
  management, light and, 88,100i
  physiology, characteristics of, 88-96,97t,98-99
  rotation, 64
Crops
  losses to disease and pests, 110
  need for alternative, 58,64
Crown gall, cause of, 121
Cucumber mosaic virus (CMV), 114,121
Cultivar variety, recommendations, 63-64
Cultivation, legal restrictions on, 8,19,57,58,61
Cultural controls, disease process management, 124
Curculios, hemp pest, 112
Curvularia cymbopogonis, hemp diseases, 120
Curvularia lunata, hemp diseases, 120
Cutworms, hemp pest, 115
  NVP for, 123
Cyst nematodes, hemp disease, 121

Damping off, 117,124
Daylength
  effects of, 2
  plant production and, 71
Decortication, 78,79
Deforestation, papermaking industry, 86
Degree of polymerization (DP), 236
Delia platura, hemp pest, 115
Delia radicum, hemp pest, 115
Delignification, hemp woody core, 233
Densitometer, TOXI-MS Cannabinoid Test, 53

Depolymerization, woody core pulp, 236
Deroceras reticulatum, hemp pest, 116
Desiccation, and cannabinoid content, 26-27
Dew retting, textile production, 74-75,76
Dewey, L. H., 153,159,160
Diatomaceous earth, biorational pesticide, 125
"Diazo" test, THC content, 175
Dichroplus maculipennis, hemp pest, 115
Dihomo-gamma-linolenic acid (DGLA), 250
Dioecious hemp, 2,15,70,153-159, 162,163,178
  selection for fiber content, 164-165,167,172
Disease control, 122-126
Ditylenchus dipsaci, hemp disease, 121
Dodder (*Cuscuta ssp.*), hemp parasite, 122
Domestication, 5-7
  base for classification, 136
Dominant characteristics, 158-159,173
Dot moth (*Melanchra persicariae*), hemp pest, 113
Downy mildew, cause of, 120
Drug biotype, diseases and pests, 109-110
Drug hemp (resin), domestication, 136
Drug strains, 147-148
Drug-type landraces, 138
Dry matter
  kenaf, 104
  partitioning of, 95
  production of, 79-80,81
Durban, South Africa, 147

Eastern Europe, growth of hemp, 86
Economics and market research, Hemp Research Programme, 87
Eletta Campana, hemp variety, 57

Emergence date, light and, 88
Empoasca fabae, hemp pest, 114
Empoasca flavescens, hemp pest, 114
Encarsia formosa, parasitoid, 123
Endocylyta excrescens, 111
Energy use, papermaking industry, 86
English sparrow, hemp pest, 116
Ensiling, use of, 91
Entomophthora thripidum, fungus, 115
"Entry holes," ECB, 111
Environmental stresses and cannabinoid production, 26-32,34
Enzyme Multiplied Immunoassay (EMIT), 52,53
Enzyme-linked immunosorbent assay (ELISA), assay format, 48-51
Epistasis, 163
Ermakovskaya Mestnaya, Siberian cultivar, 145
Escherichia coli, 47
Essential fatty acids (EFAs), 247,248
Europe, hemp and flax in, 85
European chafers (*Melolontha hippocastani*), hemp pest, 112
European Community
  GC/MS method, 54
  hemp THC content, 50,175
European corn borers (ECB), hemp pest, 110-111
  parasitoid for, 123
European fruit scale (*Parthenolecanium corni*), hemp pest, 114
Eurybrachys tomentosa, hemp pest, 114
Evening primrose (*Oenothera biennis* L.), source of GLA, 250
Ex situ, germplasm, 133. *See also* Genebank collections; Working collections, research
Explants, use of, 186,189,191i-194i
Extraction, hemp seed oil, 245
Extrusion pulping, 225,227,237
  ATO-DLO test, 227
EZ-Screen, 51

False chinch bugs (*Nysius ericae*),
    hemp pest, 114
Far Eastern hemp landraces, 146
Fedora 19, hemp variety,
    63,141,171,174
Fédrina, hemp variety,
    57,63,71,74,171,174
    bark content in, 98
    stem growth, 95-96,97t
Feeding stimulants, GRRs, 126
Felina 34, hemp variety, 63,142,
    171,172,174
Female habit, hemp, 168
Female plant
    characteristics of, 167-168
    description of, 4,21,70
    genetic research, 154-154,162,167,
        170,173t,173-174,178
Ferimon, hemp variety, 174
Ferrara, hemp variety, 160
Ferrarese, hemp variety, 62
Fertility, 69-70
Fertilizer use, cotton industry, 86
Fiber
    biotype, diseases and pests, 109-110
    bonding, 234
    landraces, original, 7,62-63
    length, woody core pulp, 234-235
    previous hemp crop, 61
    and seed strains, commercial,
        141-146
Fiber content
    genetic selection for, 159-160
    micropropagation and, 187
    selection for, 164-175
Fiber hemp
    domestication, 136
    cultivars, 16,57,62-63
    cultivation, problems related to, 61
    THC content evaluation, 56-57
Fibranova, hemp variety, 57,177,202,
    203i,206,207t
Fibrimon 56, hemp variety, 16,17,57,
    141,165,166t
    propagation, 63,189

Field density, effects on growth, 2
Field drying, 91-92
Fingerprinting, RAPD, 199
FIN-313, seed yield, 244
FIN-314, grain seed cultivar, 17
Flavescent leafhopper (*Empoasca
    flavescens*), hemp pest, 114
Flavonoids, 22
Flax (*Linum usitatissimum*)
    alternative crop, 103
    early modern use, 85
Flea beetles, hemp pests,
    110,111-112
Fleischmann, R., 153,159,160,163
Flotation, 78,79
Flowering, 3-4,70-71,93-94
Flowering date, 71,81,91
    plant density, 68
    stem growth, 98
Flowers
    caterpillar damage, hemp pest, 113
    ECB, 111
    HBs, 111
Fluorescence Polarization
    Immunoassay (TDx), 52
France, extrusion pulping, 225
French breeders, hybrid varieties,
    171,174
French cultivars, 63,141-142
Frontline, 51-52
Frost damage, 90,91,101
Fungal metabolism of THC, 31
Fungus, hemp plants, 109,115,
    116-120,123-124
Fusarium oxysporum, 109,118-119
Fusarium solani, root rot, 118,121
Fusarium species
    root rot, 118-119
    stalk cankers, 118
Futura, hemp variety, 77,63,174,189

Gamma-linolenic acid (GLA), 248,250
    sources of, 251t
Ganja landraces, Caribbean, 12

Garden tiger moth (*Arctia caja*), hemp pest, 113
Gas-chromatography (GC) analysis, 50,50i,51,53-55,58
G.A.T.E. Rudolf Fleischmann Research Institute, 16
Geisha distinctissima, hemp pest, 114
Genebank collections, 148-149
Genetic engineering, 185,208-209
Genetic erosion, 133
Genetic maps, 198
Genetic variation, importance of knowledge of, 202
Geographical provenance, 137-138
German cultivars, 146
Germination, 2,70
  temperature and, 66
Germplasm, definition, 133
Gibberella teleomorphs, 118
Glandular hair, THC content and, 176
Glasshouse leafhopper (*Zygina pallidifrons*), hemp pest, 114
Goat moth (*Cossus cossus*), 111
Gold Labeled Optically-read Rapid Immuno Assay (GLORIA), 51-52
Graphocephala coccinea, hemp pest, 114
Grapholita delineana, 110
Grasshoppers, hemp pest, 115
Gray mold, fungus (*Botrytis cinerea*), 87-88,117,124
Green peach aphid (*Myzus persicae*), hemp pest, 114
Green stink bugs (*Nezara viridula*), hemp pest, 114
Greenhouse thrips (*Heliothrips haemorrhoidalis*), hemp pest, 115
Greenhouse whitefly (*Trialeurodes vaporariorum*), hemp pest, 114
Grishko, N. N., 153-154,155

Growth and reproduction regulators (GRRs), biorational pesticide, 125-126
Growth regulators, micropropagation and, 187,188i,188t,189,190, 195-196
Grubs, hemp pest, 112
  microbial pesticides for, 123
Gryllus chinensis, hemp pest, 115
Gryllus desertus, hemp pest, 115
Gymnetron labile, hemp pest, 112
Gymnetron pascuorum, hemp pest, 112

Hallucinogenic substances, hemp cultivation and, 61
"Hanckling," 75
Handsheets, bast content, 73
Harvest and storage technology, Hemp Research Programme, 87
Harvest date
  light and, 88,89,91-93,92t,100i, 101,102i
  THC content and, 176
Harvest index (HI), crop growth equation, 88
Harvesting, textile production, 74
Hashish
  diseases and pests, 109-110
  production, 138
  THC content, 176
Helicoverpa zea, hemp pest, 113
Heliothis armigera, hemp pest, 113
Heliothis viriplaca, hemp pest, 113
Heliothrips haemorrhoidalis, hemp pest, 115
Hemicellulose, 72
Hemp
  breeding, 15-18,87
    historical review, 153-155
  cytology, early, 154
  early modern use, 85-86
  fibers, quality criteria, 76

Hemp *(continued)*
  lignin content, 86
  stem, chemical composition
    of, 72-73
Hemp blast fiber preparation,
  ATO-DLO test, 227
Hemp borers (HBs) *(Grapholita
  delineana)*, 110,111,116
Hemp flea beetles *(Psylliodes
  attenuata)*, 111,113,115
  resistance to, 161,178
Hemp linnet, hemp pest, 116
Hemp longhorn beetles *(Thyestes
  gebleri)*, hemp pest, 112
Hemp louse *(Phorodon cannabis)*,
  hemp pest, 114
Hemp mosaic virus (HMV), 114,121
Hemp Research Programme, 87
Hemp russet mite *(Aculops
  cannabicola)*, hemp pest, 116
Hemp seed oil, 243-244
  EFAs content, 247,247t,252
  fatty acids content, 247t
  GLA content, 251,251t
  trace amounts of THC in, 246
Hemp streak virus (HSV),
  114,115,121
Hemp weevil *(Rhinocus
  pericarpius)*, 112
Heteridera humuli, hemp disease, 121
Heteridera schachtii, hemp disease,
  121
Heterochromosomes, 154,155,205
Heterocyclic compounds, synthetic
  chemical pesticides, 126
"Heterosis" effect, 160,161,162
Hieroglyphus nigrorepletus, hemp
  pest, 115
High Performance Liquid
  Chromatography (HPLC),
  54-55
High UV-B exposure, and THC
  production, 31
Hindu Kush, drug strain, 147

Hop *(Humulus lupulus)*, RAPD
  analysis, 201
Hops aphid *(Phorodon humuli)*,
  hemp pest, 114
Hops cyst nematode *(Heteridera
  humuli)*, hemp disease, 121
HortaPharm BV
  cannabinoid extracts, 17,18
  working collection, 148
Horticultural oil, biorational
  pesticide, 125
Humidity, and cannabinoid content, 27
Hungarian cultivars, 63,67-68,142-143
Hybrid B-7, 160,162
"Hybrid populations," 174
Hybridization, hemp plants,
  155,158,164,171,174
Hybridoma line, 46
Hymenoscyphus herbarum, stalk
  cankers, 118

Icerya purchasi, hemp pest, 114
Idolbutirric acid (IBA), growth
  regulator, 187,189
Immuno-chromatographic assay,
  51-52
Immunogen, preparation of, 44
Immunological analysis, 44-47
Impregnation and preheating,
  ATO-DLO test, 227
In situ, germplasm, 133
Inbreeding, 154-155,163-164
India, early use for psychoactive
  properties, 8
Indiana University, working
  collection, 148
Industrial hemp, and legal
  restrictions, 19
Inflorescence, gene for, 159
Inheritance, hemp plants,
  154-155,209
Insect predation, and resins, 28-30
Insect-repellent properties terpenes, 29

Intersexual forms, types of, 168-169
Italian cultivars, 63,145,160

Japanese ghost moth (*Endocylyta excrescens*), 111
Japanese hemp, 12
Java root knot nematodes (*Meloidogyne javanica*), hemp disease, 121
Jute, hemp replacement, 86
Juvenoids, biorational pesticide, 125

Kansas, soil moisture studies, 27
Kenaf, alternative crop, 88,102-103
Kenevir, stem growth, 95,97t
Kentucky hemp cultivars, 12,146
Kinai unisexualis, stem growth, 95,97t
Kompolti, Hungarian cultivar, 63,142, 143,160,165,167,173t,174
  THC content in, 176,176-177
Kompolti Hybrid TC, 63,67-68,71
  bark content, 98
  stem growth, 95-96,97t
  "heterosis," 162-163
Kompolti Hyper Elite
  bark content, 98
  stem growth, 96,97t,101
Kompolti Sárgaszárú variety, 142,227
Kozuhara zairai
  bark content, 98
  stem growth, 96,97t
Kuban cultivar, 144
Kymington, hemp variety, 160

Lacewings (*Chrysoperla carnea*), biocontrol predator, 122-123
Landraces, 7,12,62-63
  Chinese, 153,159
Lax inflorescence, gene for, 159
Leaf appearance
  kenaf, 103
  light and, 89,100i
  plant density and, 68

Leaf appearance (*continued*)
  temperature and, 66-67,89
Leaf diseases, 31,111,119-120
Leafhoppers, hemp pests, 114
Leafminer wasps (*Dacnusua sibirica*), parasitoid, 123
Leafminers, hemp pest, 113-114
Leptosphaeria acuta, stalk cankers, 118
Leptosphaeria cannabina, 119-120
Leptosphaerulina trifolii, pepper spot, 120
Leveillula taurica, powdery mildew, 120
Licoris tripustulatus, hemp pest, 115
Life cycle, 1-5,71
Light, crop growth impact, 88-89
Light interception and utilization (LINTUL) model, 90,100i, 101,102i
Lighting oil, 8
Lignin, 72,73
  hemp stem content, 80,81
  removal, 232
  softening, 231
Limax maximus, hemp pest, 116
Linoleic acid (LA), 247,247t,248,250
Linolenic acid (LNA), 247,247t,248
  deficiency treatment, 252
Linseed oil (flax), 244,252
Liriomyza cannabis, hemp pest, 114
Liriomyza eupatorii, hemp pest, 114
Local adaptations, 137
Locusts, hemp pest, 115
Logistics chain, 76
Longtailed mealy bug (*Pseudococcus longispinus*), hemp pest, 114
Loxostege sticticalis, hemp pest, 113
Luster, hemp quality, 75
Lygus lineolaris, hemp pest, 114

Macrophomina phaseolina, fungus, 117,119
Maggots, hemp pest, 113-114,115
  microbial pesticides for, 123

Magpie, hemp pest, 116
Maize (*Zea mays*), ECBs in, 110-111
Malathion, synthetic poison, 126
Male plants, 154-159,164,167,
    171-172,205,206
  description of, 4,21,70
  stem growth, 96-97
Male-specific markers, 206-207
"Male sterile" variety, 162,164,175
Mamestra brassicae, hemp pest, 113
Mamestra configurata, hemp pest, 113
Mammals, hemp predation, 116
Marijuana
  diseases and pests, 109-110
  production, 138
Marijuana thrips (*Oxythrips cannabensis*), hemp pest, 115
Mass action law, 44
Mass Spectrometry (MS), 53-54
McPhee, H., sexual genetics, 154,155
Mealybugs, hemp pests, 114
Mechanical controls, disease process management, 124
Medisins, Netherlands drug cultivar, 148
Melanchra persicariae, hemp pest, 113
Meloidogyne chitwoodi, impact of, 64-65
Meloidogyne hapla, hemp disease, 121
  impact of, 64-65
  resistance to, 178
Meloidogyne incognita, hemp disease, 121
Meloidogyne javanica, hemp disease, 121
Melolontha melolontha, hemp pest, 112
Melolontha vulgaris, hemp pest, 112
Mendel's laws, 154
Mexican varieties, biogenetic pathways, 33

Micropropagation, 186-187,188i,188t, 189-190,191t,191i-194i,194
  THC (Δ^9-tetrahydrocannabinol) and, 187
Mildew. *See* Black mildew; Downey mildew; Pink mildew; Powdery mildew
Minerals, and cannabinoid content, 28
Minimal detectable concentration (MDC), 50
Mississippi varieties, CBC content, 32
Mites, hemp pests, 115-116
  predacious, 123
MLO. *See* Mycoplasma-like organisms (MLO)
Mobile nutrients, 122
Molecular map, 208
Molecular markers, 204-208
Monoclonal antibodies (Mabs)
  analysis base, 44,45t
  production of, 46
Monoecious hemp, 153,154-159,162, 163-164,167,172,173t,173-174, 178-179,205
  breeding for, 167-172,175
  cultivars, 5
  low-THC varieties, 16
  varieties, seed yield, 244
Mordellistena micans, hemp pest, 112
Mordellistena parvula, hemp pest, 112
Morphological markers, THC and, 62,177
Mycoplasma-like organisms (MLO), hemp disease, 121
Myzus persicae, hemp pest, 114

Naphtalenic acid (NAA), growth regulator, 187
National Institute on Drug Abuse (NIDA), GC analysis, 53
Neem, biorational pesticide, 125
Nematodes
  biocontrol with, 124
  hemp disease, 121

Netherlands CPRO, working collection, 148
"New" fibers, 102-103
Nezara viridula, hemp pest, 114
Nicotiana tabacum, 196
Nicotine, biorational pesticide, 125
Nitrogen fertilization, use of, 69-70,72
Noctuid budworms, hemp pest, 113,116
NPV for, 123
Nonhybrid inbreds, drug strains, 147
"Normal axis" method, 165-167,166t
North America, hemp and flax in, 85
Northern Lights, Washington State cross-bred cultivar, 147
Northern root knot nematodes (*Meloidogyne hapla*), hemp disease, 121
Novosadska Konoply variety, 2,3i
Nuclear polyhedrosis virus (NPV), 123
Nuthatch, hemp pest, 116
Nutrient, imbalance diseases, 122
Nysius ericae, hemp pest, 114

Odontotermes obesus, hemp pest, 112
Official Bulletin of European Communities, GC/MS method, 54
Oilseed varieties
  recent development of, 17
  previous hemp crop, 61
"Olifera," Russian hemp seed oil, 243
Olive leaf spot, cause of, 120
"One year reserve seed" method, 170
Onion thrips (*Thrips tabaci*), hemp pest, 115
Organic farming, 86
Organic pesticides, 125
Organophosphate, synthetic chemical pesticides, 126
Orobanche ramosa, root rot, 118,122,178
Ostrinia nubilalis, 110,178

Outcrossing, obligate, 15
Oxythrips cannabensis, hemp pest, 115

Pakistan, *indica* landraces, 147
Palmate leaves, 21
Panorama, Hungarian ornamental cultivar, 148
Papaipema cataphracta, 111
Papaipema nebris, 111
Paper qualities, 234-236
Papermaking
  cellulose content, 73
  Chinese invention, 230
  core content, 73
  hemp production for, 77-79,85
  industry, criticisms of, 86
  lignin content, 73
  plant density and, 67
Parasites, 122
Parasitoids, biocontrol, 123
Parthenolecanium corni, hemp pest, 114
Passenger pigeon, hemp pest, 116
Pepper spot, cause of, 120
Pest control, 122-126
"Pesticide," 125
Pesticides
  cotton industry, 86
  microbial, 123-124
Pest-tolerant, 110
Petiole, 70
Pharmaceutical varieties, recent development of, 17
Phenological development, hemp trait, 64,70-71
Phenological events, 70
Phenotypical trait, genetic map, 198
Pheromone, disease control, 125-126
Phomopsis cannabina, stalk cankers, 118
Phomopsis ganjae, white leaf spot, 120
Phorodon cannabis, hemp pest, 114
Phorodon humuli, hemp pest, 114
Photoperiod, critical, 71

Phyllotreta atra, hemp pest, 112
Phyllotreta nemorum, hemp pest, 112,114
Phytomyza horticola, hemp pest, 114
Pilaenus spumarius, hemp pest, 114
Pink mildew, cause of, 120
Pirate bug, biocontrol predator, 123
Pistacia vera, 207
Pistillate, 70
Plant breeding, Hemp Research Programme, 87
Plant density, 101
  bark content and, 98-99
  quality and, 67-69,81
  RUE, 94-95
Plant pathology, Hemp Research Programme, 87
Plant tissue culture, 185-190
Planting, methods of, 65
Plants, DNA polymorphism in, 198
Podagrica aerata, hemp pest, 112
Podagrica malvae, hemp pest, 112
Polish cultivars, 63,143
Polyclonal antibodies
  analysis base, 44,45t
  production of, 45-46
Polydrosus sericeus, hemp pest, 112
Polymerase Chain Reaction (PCR) technique, 199,201,202
Pooled sera, use of, 45
Potato bugs (*Calocoris norrvegicuss*), hemp pest, 114-115
Potato leafhopper (*Empoasca fabae*), hemp pest, 114
Powdery mildew, cause of, 120
"Premature wilt," 119
Primary dispersal, pre-Christian era, 8,9i
Printing/writing paper, APXP hemp, 229,237
Prostaglandin metabolic pathways, 248,249i
Pseudaulacaspis pentagona, hemp pest, 114

Pseudococcus longispinus, hemp pest, 114
Pseudomonas syringae, hemp disease, 121
Pseudoperonospora cannabina, downy mildew, 120
Pseudoperonospora humuli, downy mildew, 120
Psychoactive drug use, 7
Psychoactive properties, 8,86
Psychoactivity, priority over oil and protein potential, 253
Psylliodes attenuata, hemp pest, 111,113,115,161,178
Psylliodes punctulata, hemp pest, 111-112
Pulp technology, Hemp Research Programme, 87
Pyrethrum, biorational pesticide, 125
Pythium aphanidermatum, fungus, 117
Pythium ultimum, fungus, 117

Quality, textile production, 75-76

Radioactive tracer, problem of, 48
Radioimmunoassay (RIA), assay format, 47-48
Rainfall, fiber hemp requirement, 69
Ramie (*Boehmeria nivea*), 102
Random Amplified Polymorphic DNA (RAPD) technique, 199-202, 203i,204,205-206,207
  genetic diversity, 179
Recalcitrant species, hemp as, 194,208
Recombinant antibodies, production of, 46-47
Redbanded leafhopper (*Graphocephala coccinea*), hemp pest, 114
Refiner mechanical pulping (RMP), 228,229
Refining, ATO-DLO test, 228
Reproductive pheromone, biorational pesticide, 125-126

Resins
  production, 24,71
  and wounding, 28
  water loss protection, 26
Resistance, breeding for, 178
Restriction Fragment Length
    Polymorphism (RFLP)
    technique, 198-201,204,207
Retting, 74-75,76,81
  technologies, need for
    development of, 76-77,81
Reverse screw element (RSE), pulp
    extrusion, 226
Rhinocus pericarpius, hemp pest, 112
Rhizoctonia solani, 117,118
Ricania japonica, hemp pest, 114
Root knot nematodes, hemp disease,
    121
Root rot, 118
Roots
  caterpillars in, 111
  insect pests, 110-112
Row width, impact of, 72
RUE. *See* Average radiation-use
    efficiency (RUE)
Rumanian cultivars, 143
Russia, hemp and flax in, 85
Russian cultivars, 144-145
Ryania, biorational pesticide, 125

Sample, preparation for GC/MS
    method, 54-55
Santhica 23, THC-free cultivar, 142
Scachard data plot, 48
Scales, hemp pests, 114
  biocontrol of, 123
Schiffnerula cannabis, black mildew,
    120
Sclerotinia sclerotiorum, 117,124
Screening kits, one-step, 51
"Scutching," 75
Seasonal variation, hemp content, 73
Secondary bast fiber, quality
    criterion, 76

Secondary dispersal, historical
    period, 10i,12
Secondary metabolites, 185,194-195
Seed landraces, 136
Seed maturity, 71
Seed treatments, 126
Seed yield, 162,174,178-179
  inbreeding and, 163
Seedbed, hemp, 65
Seeding rate, 67-69,79
Seedlings, hemp pests, 115
Seeds
  biocontrols for, 124,126
  caterpillar damage, hemp pest, 113
  germination, 2
  HBs, 111
  size accessions, 5,6i
  use of, 7,85
Selective breeding, difficulties
    of, 15,16
Selective predators, 123
Self-thinning
  danger of, 67-68
  row width impact, 72
  RUE, 93,94-95
Sensitivity
  competitive ELISA, 49
  GC/MS, 54
Septoria cannabis, leaf disease, 119
Septoria neocannabina, leaf disease,
    119
Sessile epidermal glands, 24
Sex determination, markers and,
    204-208
Sexual dimorphism, hemp, 167
Sexual genetics, hemp plants,
    154-159
Shoot culture technique, 186,190,
    191t,191i-194i
Shoot regeneration, 194
Short-span compressive test (SCT),
    linerboard, 229
Sieving, 78,79
Silver Y-mouth (*Autographa
    gamma*), hemp pest, 113

Silverleaf white fly (*Bemisia argentfolii*), hemp pest, 114
Sitona ssp., hemp pest, 112
Skunk #1, California cross-bred cultivar, 147
Slugs, hemp pests, 116
Sodium hydroxide (NaOH), 223,231,233,236
Soil fumigants, 64
Soil nitrogen, bark content, 99
Soil nutrients, and cannabinoid content, 28
Soil pathogens
  hemp impact on, 64-65
  resistance, hemp trait, 64
Solenopsis geminata, hemp pest, 112
"Sore skin," 118
Southern blight, cause of, 119
Southern root knot nematodes (*Meloidogyne incognita*), hemp disease, 121
Soviet Union
  cultivars, 63
  growth of hemp, 86
Sowing date, light and, 88,89t,89-91, 91t,92t,100i,101,102i
Specificity, recombinant antibodies, 47
Sphaeria cannabis, stalk cankers, 118
Sphaerotheca macularis, powdery mildew, 120
Spider mites, hemp pests, 110,115-116
  biocontrol of, 123
Spirulina, source of GLA, 250
Spittlebug (*Pilaenus spumarius*), hemp pest, 114
Spodoptera exigua, hemp pest, 113,115
Spodoptera litura, hemp pest, 115
Sprinkled locust (*Chloealtis conspersa*), hemp pest, 115
Stalk cankers, fungus, 117-118
Stalked epidermal glands, 24,25i
Stalks
  development of, 70-71
  insect pests, 110-112

Staminate, 70
Starlings, hemp pest, 116
Stearidonic acid (SDA), 248,250,251-252
Stem
  development of, 70-71
  dry matter, 95-96,97t,101,102i
  bark content in, 98-99
  fasciation, inbreeding, 163
  quality, hemp trait, 64
Stem borers, hemp pests, 110
Stem cutting rooting, marijuana growing, 16
Stem nematode (*Ditylenchus dipsaci*), hemp disease, 121
Stem yield
  fiber, selection for, 165,166t,167
  inbreeding's impact on, 163-164
  selection for, 159,160
Stemphylium leaf spot, cause of, 120
Stenocranus quiandainus, hemp pest, 114
Straw pulp, 235,237
Subdioicus plant, 154
Sugar beet, 89-90
Sugar beet cyst nematode (*Heteridera schachtii*), hemp disease, 121
Superfibra, hemp variety, 57
SureStep kit, 51
Sweet potato whitefly (*Bemisia tabaci*), hemp pest, 114
Synthetic fabrics, hemp replacement, 62,86
Synthetic pesticides, chemical, 126
Synthetic pheromone, disease control, 125

Tarnished plant bugs (*Lygus lineolaris*), hemp pest, 114
Taxonomy, 13-14,135-136
Tear strength
  and fiber length, 234
  woody core pulp, 236

Temperature
  and cannabinoid content, 27-28
  kenaf, 103
  role in hemp growth, 66-67,89i, 89-91
Tensile and burst strength, woody core pulp, 234,235
Termites, hemp pests, 112
Terpenes, 22,26,28
  competition suppressant action, 30
Tertiary dispersal, modern period, 11i,12
Testliner, AXP hemp, 229,237
Tetrahydrocannabivarina, cannabinoid type, 44
Tetranychus cinnabarinus, hemp pest, 115-116
Tetranychus urticae, hemp pest, 115-116
Tettigonia cantans, hemp pest, 115
Textiles, hemp production for, 8,74-77,85
THC (Δ^8-tetrahydrocannabinol), drug characteristics, 43,43i,45t
THC (Δ^9-tetrahydrocannabinol), 22,23i,26
  breeding for reduced, 175-177
  drug characteristics, 43,43i,45t,45i
  gene identification of, 208
  landraces, 7
  legal limit, 50,175
Thermal time, 66-67
Thermo-mechanical pulp (TMP) treatment, 228,229
Thin Layer Chromatography (TLC), 53
Thrips, hemp pests, 114,115
  biocontrol of, 123
Thrips tabaci, hemp pest, 115
Thrips wasps (*Thripobius semileuteus*), parasitoid, 123
Thyestes gebleri, hemp pest, 112
Tobacco whitefly (*Bemisia tabaci*), hemp pest, 114
Toscana, hemp variety, 62

TOXI-MS Cannabinoid Test, Toxi Lab Inc., 53
Tree plantations, papermaking industry, 86
Tree sparrow, hemp pest, 116
Trialeurodes vaporariorum, hemp pest, 114
Trichothecium roseum, pink mildew, 120
Tropical strains, 33
Turtledove, hemp pest, 116
Twig blight, cause of, 119

Ukrainian Research Institute of Bast Crops, working collection, 148
Ultraviolet radiation
  plant stress, 31-32
  selection pressure, and high THC/CBC varieties, 34
Unbleached alkaline-mechanical pulp (AXP), 227,230
Unbleached mechanical pulp (XP), 227,228,230
UNIKO-B, unisexual hemp, 157,174
  seed yield, 244
Unisexual hemp, breeding for, 172-175,173t
Unweighted Pair Group Method with Arithmetical Analysis (UPGMA), 206
Urine, THC detection, 47,51,52,53
UV-absorption, woody core pulping, 232

Vegetation period, 159,160,161, 163,168
Verticillium dahliae, hemp impact on, 64-65
Verticillium species, root rot, 118,119,123
Viruses, hemp disease, 121,186

Wasps, parasitoids, 123
Water retting, textile production, 74
Weeds, hemp suppression of, 69,87
Weevils, hemp pest, 112
Wet harvesting method, papermaking, 78
White leaf spot, cause of, 120,121
White peach scale (*Pseudaulacaspis pentagona*), hemp pest, 114
Whiteflies, hemp pest, 114
  biocontrol of, 123
Whitefly wasp (*Encarsia formosa*), parasitoid, 123
Whole hemp seed, nutrient content, 243
Wild *Cannabis*, Central Asia, 137
Wind pollination, 3,15
Woodfree printing paper, 226
Woodpecker, hemp pest, 116
Working collections, research, 148

Xanthomonas campestris, hemp disease, 121
Xylem fiber length, quality criterion, 76

Yellow leaf spot, cause of, 119
Yield
  equation for, 88
  papermaking, 77
  potential, 88,101
  textile production, 75

Zea mays, 110-111
Zygina pallidifrons, hemp pest, 114

Order Your Own Copy of This Important Book for Your Personal Library!

ADVANCES IN HEMP RESEARCH

_____ in hardbound at $69.95 (ISBN: 1-56022-872-5)

COST OF BOOKS _____	☐ **BILL ME LATER:** ($5 service charge will be added) (Bill-me option is good on US/Canada/Mexico orders only; not good to jobbers, wholesalers, or subscription agencies.)
OUTSIDE USA/CANADA/ MEXICO: ADD 20% _____	
POSTAGE & HANDLING _____ (US: $3.00 for first book & $1.25 for each additional book) Outside US: $4.75 for first book & $1.75 for each additional book)	☐ Check here if billing address is different from shipping address and attach purchase order and billing address information.
	Signature _____
SUBTOTAL _____	☐ **PAYMENT ENCLOSED:** $ _____
IN CANADA: ADD 7% GST _____	☐ **PLEASE CHARGE TO MY CREDIT CARD.**
STATE TAX _____ (NY, OH & MN residents, please add appropriate local sales tax)	☐ Visa ☐ MasterCard ☐ AmEx ☐ Discover
	Account # _____
FINAL TOTAL _____ (If paying in Canadian funds, convert using the current exchange rate. UNESCO coupons welcome.)	Exp. Date _____
	Signature _____

Prices in US dollars and subject to change without notice.

NAME _____

INSTITUTION _____

ADDRESS _____

CITY _____

STATE/ZIP _____

COUNTRY _____ COUNTY (NY residents only) _____

TEL _____ FAX _____

E-MAIL _____

May we use your e-mail address for confirmations and other types of information? ☐ Yes ☐ No

Order From Your Local Bookstore or Directly From

The Haworth Press, Inc.

10 Alice Street, Binghamton, New York 13904-1580 • USA

TELEPHONE: 1-800-HAWORTH (1-800-429-6784) / Outside US/Canada: (607) 722-5857

FAX: 1-800-895-0582 / Outside US/Canada: (607) 772-6362

E-mail: getinfo@haworthpressinc.com

PLEASE PHOTOCOPY THIS FORM FOR YOUR PERSONAL USE.

2012 046